APPLICATIONS OF OPTICAL DIGITAL DATA DISK STORAGE SYSTEMS

Volume 490

Contents

Conference Committee

Chairmen: **W. Mike Deese,** Cherokee Data Systems, USA
Marino Carasso, Philips Research Laboratories, The Netherlands

Organizing Committee

Gentil Verbeke, Agfa-Gevaert N. V., Belgium; **Lionel Baker,** Sira Ltd., England

Program Committee

Mark Goldberg, Department of Defense, USA
Francois LeCarvennec, Thomson-CSF France
Masafumi Mori, Toshiba R&D Center, Japan

Belgian Organizing Host Committee

Jean Ebbeni, Université Libre de Bruxelles, Belgium
André Monfils, Université de Liège, Belgium

International Activities Committee Chairman

Andrew G. Tescher, The Aerospace Corporation, USA

Session Chairmen

Session 1—Optical Digital Data Systems, **Michel S. Thomas,** Thomson-CSF, France
Session 2—Applications, **Ann U. Kerr,** DARPA, USA
Session 3—Standardization, **J. A. Th. Verhoeven,** Philips Research Laboratories,
　　　The Netherlands; **Joseph Zajaczkowski,** Cherokee Data Systems, USA
Session 4—Future Directions, **Masafumi Mori,** Toshiba R&D Center, Japan;
　　　Mark Goldberg, Department of Defense, USA

INTRODUCTION

The mission of the Brussels conference was to bring together the developers and potential users of systems which employ optically recorded digital information. This was a users' conference and focused mainly on applications of optical digital data storage systems, their capabilities, uses, and future.

Within the next year, the commercial availability of both write-once and read-only optical disk storage systems with gigabyte capacities will create vast new opportunities for the high performance, low-cost storage of machine-readable information. Optical storage technologies have moved out of the research labs and are now emerging as products in the commercial marketplace. While the developers of these emerging products have carefully studied their vast applications, many of the potential users are only recently coming to understand the exciting advantages which optical storage offers over the more conventional forms of data storage.

The 1984 conference was carefully planned to offer a global view of optical digital data disk storage systems—the history of their technical development, their current and future applications, and what their future technical growth and competition might be. A special session covering the need for and pursuit of optical data interchange standards was also included.

W. Mike Deese
Cherokee Data Systems, USA

Marino Carasso
Philips Research Laboratories, The Netherlands

APPLICATIONS OF OPTICAL DIGITAL DATA DISK STORAGE SYSTEMS

Volume 490

Session 1

Optical Digital Data Systems

Chairman
Michel S. Thomas
Thomson-CSF, France

The optical digital data disk for the storage of images of original documents

A.J.L.R. Hasekamp, W. Hoekstra, J.W. Klimbie

Philips Data Systems
P.O.B. 245, 7300 AE Apeldoorn, The Netherlands

Abstract

Based on the Digital Optical Recorder Philips has developed a computer system, called "Megadoc", for the storage and retrieval of several millions of images of original documents to be used in office automation applications. The design features modularity and expandibilaty to allow growth and easy interfacing (one to many workstations, limited to vast storage capacity), applying as much as possible normal EDP components and peripherals.
Standard EDP and WP are also available in Megadoc, such that the storage of EDP, WP and imaginal data can be intermixed.

Optical recording in general

The idea to use the reflected light of a laser beam to detect "irregularaties" - forming patterns of information - on a surface emerged some 15 years ago.

Based on this principle of optical recording Philips developed three different product types:
a. The compact disc and dito player for the hifi consumer audio market (the same principle can be used to distribute alphanumeric information, if the player is connected to a computer).
b. The video long play or laser vision disc and player intended for TV-like programmes, already introduced on the US-market some years ago, and recently in U.K. and other Western European countries.
   (Note that both products are replay-only types).
c. The digital optical disc and recorder, intended for the storage of large quantities of computer oriented data, introduced on the EDP-market in 1984.
   (The user buys a virgin disc, and the recorder writes and read onto/ from this disc).

This paper deals with digital optical recording and its applications: in particular the field of electronic document storage and delivery has a vast potential.

Principles of digital optical recording

The DOR disc is formed from two 12 inch glass substrates placed back to back separated by outer and inner metal rings. In this so-called sandwich on both inner sides a very thin film of a tellurium alloy is deposited, and a spiral groove structure is pressed into this layer.
The spiral groove can be regarded as the equivalent of 32,000 usable tracks, and every track is divided into 32 segments.
The segment is the basic unit of processing on the DOR disc. The user can write his data onto such a segment (upto 1024 user bytes) and he can address every segment individually.
Every segment is preceded by its own address. During the manufacturing process these address headings are also pressed into the tellurium layer.
The disc, placed on a sophisticated turn table, rotates 8 revolutions per second. On the access arm, which is similar in construction to the access arm of a magnetic drive, a diode laser with a complicated lense system is mounted.
Roughly speaking writing of the data into the tellurium is done by intensifying the light output of the laser beam for every "1" bit that has to be stored.
The heat caused by the light focussed on the lowest of the two layers melts a hole in the tellurium alloy (one of reasons for chosing tellurium is that it has a relatively low melting-point).
The holes are less than one micron in diameter, allowing a very high bit density (in fact a factor of 10 higher than current magnetic discs).

The reading of the written data is done by the same laser mechanism. From the light that is reflected by the track, the previously written bit pattern can be derived.

The physical process of melting holes cannot be reversed, so information once written onto the disc cannot be changed afterwards.
The tellurium alloy is stable enough to enable reading of data 10 years after the event of writing.
So, the DOR disc is a high volume (10^9 user bytes per disc side), write once, random access (mean access time 200 ms.) archival quality storage medium.

Looking from the computer point of view a DOR disc drive is a peripheral treated by the computer software in, roughly speaking, the same way as a magnetic disc drive. Consequently a DOR disc drive is always a part of a computer configuration, and by its design, it is intended to store bitstreams which have been processed by a computer.

These bitstreams may represent:
a. Information coded by humans (e.g. via a wordprocessor).
b. Information read automatically by an optical character reader.
c. Data produced by a computer.
d. Data produced by measurement equipment.
e. Information that is scanned and digitized electronically, similar to the facsimile process.
Categories a, b and c are discussed in more detail in the next paragraph.
Measurement equipment (e.g. medical (tomoscan, X-ray), seismic, satelites) produces enormous amounts of already digitized data, which has to be stored for later processing. The DOR is a feasible solution for the storage problem arising in these special cases.
The last category in particular needs some more explanation, as we think that the future of DOR is mainly in this area, often called "Electronic Document Storage", see the paragraphes about filing of other data, electronic document storage, office automation and potential application areas.

Filing of coded data

In the categories a, b and c as mentioned above every alphanumerical character on the source document is coded and next stored in the ASCII or EBCDIC convention. This is either the result of a conversion process performed by humans (via wordprocessor, teletex, other keyboard based hardware), or performed automatically by optical character recognition hardware (e.g. reading of banking cheques), or it is the result of some form of computer processing (e.g. invoices, banking transactions, system log).

In quite a number of EDP applications one may observe that information is subjected to a kind of life cycle phenomenon:
a. Created.
b. Captured or converted into the computer.
c. Processed.
d. Corrected.
e. Frozen or consolidated.
f. Archived.
g. Destroyed.

During capturing, processing and correction the data will be retrieved and changed very frequently: consequently during these periods of the life cycle this data will be stored on magnetic disc, a medium that is very suited for high access and change rates.
As soon as data is frozen, no changes will occur anymore and the retrieval frequency diminishes very rapidly. Some applications require retrievability of data even after 3 - 10 years after the moment of consolidation. To keep data on magnetic disc during such a period is nearly impossible:
a. Such a huge amount of magnetic data sets are often not supported by operating systems.
b. The magnetic stored data degrades in quality, refreshment at yearly or bi-yearly intervals is necessary.
c. It is expensive.
Digital optical recording is extremely well suited for this kind of application.

Filing of other data

Of course only a fraction of the data that is processed in an office can be treated in the way described in paragraph 2.
Other solutions have to be found for graphs, handwritten letters, documents for which the original lay-out and fonds have to be retained, etc.
For these kind of applications the techniques of electronic scanning digitizing and storing will become feasible in the coming years.
Every page of a document (documents can be insurance policies, incoming letters, technical drawings, newspaper clippings, articles from periodicals, etc.) is scanned in a horizontal and vertical direction with a resolution of 200 lines per inch. An A4 page in this case is divided into 2287 lines, each line having 1728 picture elements, being in total nearly 4 million "pixels" (for other sizes, and other resolution different quantities apply).
During the scanning process the hardware decides pixel for pixel whether it is "black" or "white", and generates then the electronic image, consisting of 4 M bits.
It is possible to compress such a bitstream of 4M bits by applying a special purpose encoding Such an algorithm may yield a bitstream of 5 - 10% of the original length. This compressed image can then be stored onto a peripheral computer memory like DOR.
On one side of a DOR disc one is able to store approx. 25,000 compressed images. For applica-

tions with very large requirements a jukebox is developed. In such a facility 64 DOR discs ca
be stored. Random access to the information on any disc is possible within 20 seconds. One ca
of course use several jukeboxes, thus giving access to approx. 8 - 10 million images.

Electronic document storage: Megadoc systems

Documents stored in an archive can be a combination of any of the following
a. Information produced by word or text processing system.
b. Handwritten documents, signatures.
c. Pre-printed documents.
d. Graphs, letter headings etc.
e. Computer produced data.

Until now there has been no feasible means of storing these kinds of documents in electronic form and consequently there has been no feasible way of implementing a central departmen-
tal electronic archive.
Using the digital optical recording technology this electronic document storage based on full
documents will become a reality in the next few years. We, at Philips call these systems Megadoc (Mega is million, doc. for document).
In these Megadoc systems electronic scanners are used for data capturing by transforming
every page of a document into a bitstream: very high resolution cathode-ray tubes are used
for displaying a soft copy of a page, while image printers can provide hardcopies.
For the transport of the bitstream to and from these various computer peripherals a small business computer will be used.

Depending on the application, a simple or more sophisticated retrieval software with thesauri and indices is required to enable the user to capture and to retrieve documents. Sometimes organizations can use their already implemented retrieval systems if it is possible to
adapt the interface to the new Megadoc situation. Consequently the large investments involved
in building up thesauri and citations etc. are saved.

A typical Megadoc configuration will consist of:
a. 1 or 2 input workstations (an electronic scanner, a high resolution display and a keyboard
   with visual display for the identification of the image to the system).
b. 3 to 6 retrieval workstations (a high resolution display and a keyboard with visual display to command the retrieval procedure and to visualize the image).
c. 1 or 2 image printers to provide hard copies.
d. 1 or 2 jukeboxes with optical discs, and some digital optical recorders.
e. A small business computer (processor, memory, magnetic discs).
f. Seen from the software side among else a system for the administration of all stored and
   retrievable documents is required. This administration will always reflect the requirements as set by the user.
The compression and decompression is performed in the input, respectively the output peripherals.

Office automation

Office Automation consists of applying various computerbased technologies like storage,
processing and distribution on any kind of communication between office-workers: voice, key
boarded text, compterized data, "normal" documents.
In the last years an enormous proliferation of text processing systems has occurred with the
keyboard as the major input medium. However, uptil now no feasible computerized means for
handling originals was available, although the majority of the information flow (and certainly that with the outside world) is based on non-keyboardable documents.
In this sense Megadoc, because of its image storage and distribution capabilities, can be applied as an office automation tool. Especially when large numbers of various kinds of
documents have to be processed in the office, Megadoc together with its powerful information
storage and retrieval software will become a cornerstone in this application area. Another
feature of Megadoc is that integration with other office automation tools like text processing and communication networks is easy because all are based on the same computer architecture and interfaces.
In this way Megadoc can be a milestone in any office automation growth path.

Potential application areas

In the user community one can observe a strong tendency to apply the same infrastructure
for all information processing (as well voice, as text, data and image) and to avoid the ne-
cessity that a professional worker must handle more than one terminal on his desk.

Consequently, in the long run electronic document storage and retrieval systems will, certainly in office automation environments, always be applied in combination with "normal" EDP
and text processing.

Especially in those environments where efficiency of paper handling in today's situation is a critical success factor for the enterprise, one might expect good opportunities for implementing office automation system with electronic document handling capabilities. These will be found in the following areas:

a. Banks.
   - Payment documents
   - Customer correspondence
b. Insurance Companies.
   - Policies
   - Customer files
   - Reference documents
c. Medical.
   - Patient records
   - Diagnostic images
d. Manufacturing.
   - Engineering drawings
   - Vendor/ customer records
e. Publishing, research.
   - Document storage and document delivery
f. Government.
   - Tax forms
   - Social security records
   - Engineering data
   - Land registry records
   - Personnel files
   - Libraries
   - Patent files
   - Payment records
g. Legal.
   - Reference documents
   - Case histories
h. Architectural.
   - Engineering drawings
i. Utilities.
   - Customer/ vendor records
   - Payment documents

Conclusion

Compared to other means of archival storage (paper or microform) the obvious advantages of Megadoc are:
a. Real time random access to several million images of pages.
b. Small space requirements.
c. High quality hard copies.
d. Central archiving with decentralized retrieval possibilities.
e. Storage of images as well as ASCII-coded documents.
f. Retention time of 10 years and more.
g. Integration with EDP and WP.
h. People in different places can retrieve the same document simultaneously.

High performance optical disk memory system

[1]Masafumi Mori, [2]Kazuhiko Yamamoto

[1] Toshiba Research and Development Center
1 Komukai-Toshibacho, Saiwaiku, Kawasaki, 210 Japan
[2] Toshiba Yanagicho Works
70 Yanagicho, Saiwaiku, Kawasaki, 210 Japan

Abstract

A large capacity optical disk document filing system has been developed. The system is named TOSFILE3200. One optical cassette disk can contain up to 60,000 A4 documents, and each document can be accessed in a few second. The system can also handle A3 documents in 16 dots/mm resolution suitable for the filing of drawings. The filing structure is like a familiar paper file, and it is easy to understand and remember. The aimed document can be retrieved quickly by various approaches.

Introduction

Recently the optical disk memory is attracting attention for its large capacity and fast access time.[1] Toshiba introduced the first commercial model of the optical disk memory in 1982 in Japan. It is TOSFILE2100, and is now also available in Europe. The relatively conservative specifications of our optical disk memory is to guarantee high reliability to the user, but its capability to store 10,000 documents and rapid access has great advantage over a magnetic tape memory system or a microfilm file system. It is being used in many applications such as files for consumer inquiry, real estate drawings, maps for fire fighting and consumer information.

We have now developed a high performance optical disk memory system named TOSFILE3200. This system is designed to have a much larger capacity for document filing and improved usability. In this paper we will first describe the TOSFILE3200 system generally, then the optical disk memory unit and finally the filing method.

System description of TOSFILE3200

TOSFILE3200 is an optical disk memory system which has a large capacity of 60,000 A4 documents on one double-sided disk, and up to 8 optical disk drives can be installed. It can also handle an A3 size document with 16 dots/mm resolution, which is suitable for filling drawings.

Fig. 1 shows the TOSFILE3200 system. Fig. 2 shows the system block diagram. The system consists of an optical disk memory unit, a CCD document scanner, a laser printer and a control unit including a display and a keyboard.

The resolution for reading and printing documents can be selected from either 8 dots/mm or 16 dots/mm corresponding to normal documents or drawings. A whole A4 document is bit-mapped on a 15" vertical high resolution display in 964 x 1228 dots resolution. Enlargement, revolution and scrolling operations of the document can be done.

The control unit has an ability of processing Japanese characters "KANJI". This capability is crucial in the office environment in Japan.

The system specifications are given in Table 1.

Optical disk memory unit

The optical disk memory unit consists of one to eight optical disk drives. This drive has its own controller in it to control the data format or error correction. The optical disk drive is shown in Fig. 3, and its specifications are given in Table 2.

Optical disk

A single side of the 12 inch optical disk can record 1.8G bytes of data. This is equivalent to approximately 60 magnetic tapes (30 MB, 2400 feet). The optical disk is enclosed in a plastic cassette which protects it from damage and contamination and assures high reliability of the recorded data. The optical disk is loaded automatically by inserting the cassette into the opening in front of the disk drive.

Fig. 4 shows a cassette of optical disk. Te-C is used as the recording media. Te-C has characteristics of both high sensitivity and long life.[2] Fig. 5 shows an SEM photograph of the recorded pits on the Te-C media.

The pregrooved disk has 45,000 spiral tracks and the data area is divided into blocks. Each block has a preformatted header and a blank area for 4 K Bytes of data. There are 440,000 blocks on a single side of a disk.

Table 3 shows the specifications of the optical disk.

Control and interface

The optical disk drive consists of an optical head, a positioner, a rotation system including a disk loading/unloading mechanism, and electronics.

The electronics portion does the following operations.
(a) tracking/focusing control
(b) laser beam power control and signal detection
(c) access control using a position sensor and linear motor
(d) MCLV (modified constant linear velocity) rotation control ... tracks are divided into groups and CLV control is applied group by group and CAV (constant angular velocity) control within a group
(e) signal processing ... adds a block header code and an error correcting code to information data and then modulates it when writing, demodulates a detected signal, separates a data block then detects/corrects errors, and sends data to the interface when reading.

The optical disk drive connects with a host system through a standard GPIB (IEEE 488) interface. Fig. 6 shows the block diagram of the interface. There are two data block buffers which are used alternately, one for write/read data of the optical disk and the other for communication with the host system. The effective write/read data rate is about 110 KBytes/sec, so the transfer rate on the GPIB bus should be higher than 110 KBytes/sec to prevent waiting for another disk rotation cycle.

Data format

The record format on the optical disk is controlled by the signal processor of the optical disk drive.

The data area is divided into fixed length blocks by preformatting a block pre-header at the beginning of each block. The block pre-header consists of a PLO sync code, a block sync code, block number, track number and check data. The number of the most inner block is "0", then each block is numbered serially. A data block is accessed by the block number.

Fig. 7 shows the data format. The header structure is the same as the block pre-header. The formatted data is generated by adding a sync code and an error correcting code as shown in Fig. 8.

Document filling

Filing

The file structure of TOSFILE3200 follows the concept of paper files in the office. It is easy to understand and remember. A single-side optical disk is conceptually equal to a cabinet, and it is further divided into a hierarchy of binders, documents, pages, and revisions.

Retrieval

The aimed document can be retrieved quickly by various methods, that is, direct retrieval by a registered number, multi-term retrieval using key-words, retrieval using a comment table, retrieval by the "book marker" method, and so on. This is shown in Fig. 9.

Summary

An improved document filing system has been developed. It was designed to make a system which has large capacity, meets various kinds of documents and is easy to use. These targets were fulfilled by increasing the packing density and the signal reliability of the optical disk drive, increasing the pixel resolution of the scanner and the printer, and developing user-friendly software.

It will make a great contribution in improving the efficiency of office activities.

References

1. Teruo Murakami, Isao Hoshino, Masafumi Mori, "Optical disk memory system" Proc. SPIE, vol. 329, 1982.
2. Masao Mashita, Nobuaki Yasuda, "Amorphous Te-C films for an optical disk" Proc. SPIE, vol. 329, 1982.

Table 1.   Specifications of TOSFILE3200 System

Scanner	original	max A3, sheet or book
	reading method	CCD, flat bed, document feeder
	resolution	16×15.4 dots/mm^2 or 8×7.7 dots/mm^2
	scanning speed	3 sec/A4 (at 8×7.7 dots/mm resolution)
Printer	paper	max A3, plain paper (cut sheet)
	printing method	laser scanning and electrophotography
	paper feed	double cassettes
	resolution	16×15.4 dots/mm^2
	printing speed	first print 20 sec, continuous 5 sec
Display	CRT	15", vertical, monochrome
	resolution	964 x 1228
	display mode	character, image, enlargement, inverse, revolution, scrolling
Optical disk	recording media	double sided 30 cm dia. optical disk
	recording method	DRAW using a laser light
	capacity	60,000/A4
	head access	< 0.5 sec
Interface		RS 232C, IEEE 488
Characters		alpha, numeric, kana, kanji
File Structure		cabinet > binder > document > page > revision
Record		normal, documents first, titles first
Retrieval		direct, multi-terms (key words), comment table, book maker

Table 2.   Specifications of the optical disk drive

Data transfer rate	4 M bits/sec (burst), 880 k bits/sec (continuous)
Random access speed	2.0 sec (average)
Latency time	0.3 sec (average)
Data error rate	$< 10^{-12}$
Record/retrieve data unit	4 K bytes
Interface	GPIB (IEEE 488)
MTBF	Approx. 4,000 hours
External dimensions	500 W x 230 H x 500 D
Weight	35 kg
Power consumption	200 VA (operating), 130 VA (standby)
Temperature	$10 - 35°C$ (operating), $0 - 40°C$ (non-operating)

Table 3.   Specification of the optical disk

User data capacity	1.8 G bytes/side (both side available), pre-formatted
Data track pitch	Approx. 1.6 μm
Number of tracks	45,000 tracks/side
Number of data blocks	440,000 blocks/side
Data capacity/block	4 K bytes/block
Projected life	10 years
Disk diameter	12 inches
Cassette size	340 W x 18 H x 350 D
Weight	1.3 kg
Storage conditions	Temperature $10 - 35°C$, Humidity $40 - 60 \%$ RH

Figure 1.   TOSFILE 3200 System

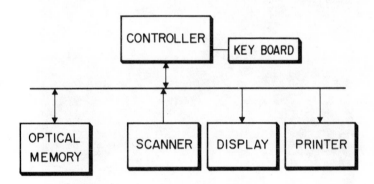

Figure 2.   System blockdiagram

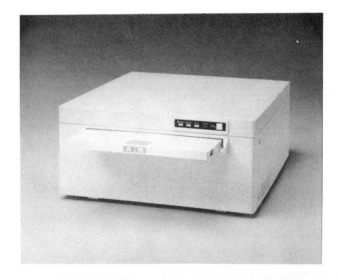

Figure 3.   Optical disk drive

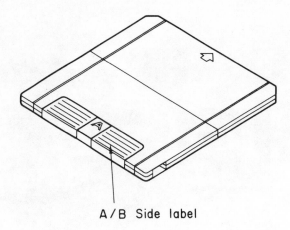

A / B Side label

Figure 4.   Optical disk cassette

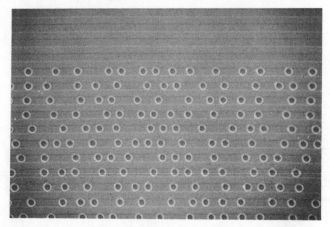

Figure 5.   Pits on Te-C optical disk

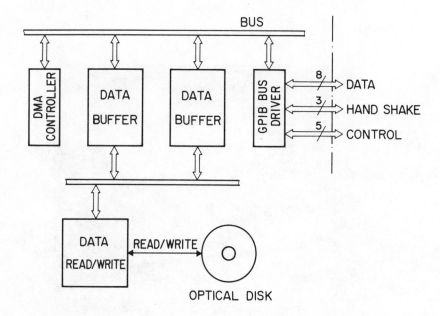

Figure 6.   Block diagram of the interface

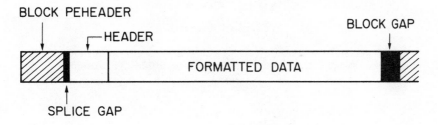

Figure 7. Data format

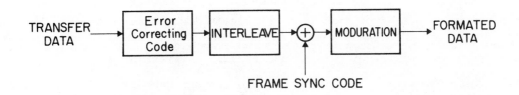

Figure 8. Formated data formation

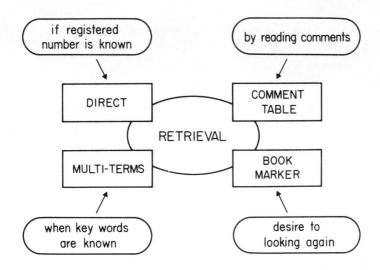

Figure 9. Retrieval methods

Development of large capacity optical disk

Yasunori Kanazawa

Hitachi Maxell Ltd. Tsukuba Research Center,
Onogo, Mitsukaido, Ibaraki, 300-25, Japan

Masaru Ito

Central Research Laboratory, Hitachi Ltd,,
Koigakubo, Kokubunji, Tokyo, 185, Japan

Shin-ichi Abe

Odawara Works, Hitachi Ltd., Kohzu, Kanagawa, 256, Japan

Abstract

A high performance DRAW type disk is developed, and the disk is used in Hitachi's newly developed file memory system, HITFILE 60. The total capacity of the system is 83 gigabytes, and the mean access time is about 6 seconds, on single disk, the mean access time is 250 msec. The resolution of the picture is about 16x15.4 lines/mm.

The disk, which diameter is 300 mmϕ, has the storage capacity is 1.3 gigabytes for single sided type (OC-301-1) or 2.6 gigabytes for double sided type (OC-301-2).The track pitch is 1.6μ m and 48,000 tracks are engraved spirally. A track is devided into 64 sectors and 512 bites of user's data can be recorded in each sector. Low error rate of $10-12$, has been confirmed after error correction, with above mentioned file memory system. Long life, more than 10 years in normal room condition, has confirmed by accellerated life test under the condition of 60°C 95 %RH.

The disk is consisted of three layers on chemically tempered glass plate with 1.1 mm thickness. The first layer is UV-curable polymer layer, the 2nd is intermediate layer of cellulose nitrate, this ensures precise pits formation on grooves, the 3rd isTe-Se-Pb alloy recording layer. The disk has the air sandwitch structure, in which two disks above mentioned are pasted together with 0.6 mm air gap.

The disk is completly protected from dust, finger prints and scratches by newly developed plastic cartridge.

Introduction

DRAW type optical disk is the most promising recording medium as a large capacity storage device.[1)] A new optical disk has been developed for HITACHI's file memory system, HITFILE 60.
The fundamental requirements for the computor use are low error rate, long term stability and high sensitivity to the write laser. For image recording, the error rate of $10^{-6} \sim 10^{-8}$ after error correction can be allowed . For computer memory use, however, the error rate has to be less than 10^{-12} after error correction. To satisfy these requirements, an air sandwitch structure using tempered glass plates are adopted. In order to preserve high density for a long time, the disk surface must be kept clean and the disk must be protected from the outer shock. A plastic cartridge protects the disk and this makes its handling easy.
Fundamental constructions of the file memorysystem are shownin Fig.1. There are two types of constructions, one is for small scale system (diffused type), and the other is for large scale system (centralized typr). In the latter application, the ODD (Optical Disk Drive) is connected to a large central processor system through OSC (Optical String Switch Controller) and ODC (Optical Disk Controller). In the former application, the ODD is connected to a small business computer through OFC (Optical Formatter Controller). These constructions are for code data recording. The system can also be used for image data recording by utilizing scanners, printers and display devices. Table 1 shows the fundamental specifications for this system.

STRUCTURE OF DISK AND ITS DATA AREA

As shown in Fig.2, the air sandwitch construction is adopted. Base plates (300 mm ϕ) are composed of chemically tempered glass. Recording layers are protected by these two glass plates. Those are pasted together with outer/inner spacers in between. Fig.3 is a photograph of this optical disk. Each recording plate has 48,000 tracks engraved spirally, and each track divided into 64 sectors.62 sectors of them are used for ordinary use and the rest are for alternate sectors. More over, in the inner region, there are 128 alternate tracks.

Data format on the recording surface is shown in Fig.4. Here, inner and outer test tracks are for the makers testing, which are not open to the users. At the head of each sector, sector informations such as sector number, track number and synchronizing marks are preformatted. Fig.5 shows data arrangement in asector. One sector is consists of 690 bytes, 512 bytes of which are for users. The dimensions of the disk are shown in Table 2.

RECORDING LAYER

The recording layer is composed of three layers as shown in Fig.6. The top layer is of Te-Se-Pb alloy in which informationpits are formed by ablation using the semiconductor laser light spot. The intermediate layer is of cellulose nitrate, this ensures precise pit formation and high signal-to-noise ratio. The bottom layer facing to the glass plate is UV-curable polymer. Fig.7 shows SEM image of the pits and grooves.

In order to improve antioxidization property, Se is added to it. However, the addition of Se gives rise to increase in grain size, which reduces signal-to-noise ratio. To suppress the grain growth, small amount of Pb is added.

Raw error rate of 10^{-6} is obtained by manufacturing the disk illuatrated above under the highly cleaned environment. This errorrate can be improved up to 10^{-12} through error correction.

The accellerated life test was carried out under the condition of 60°C, 95% RH. No performance change was observed for more than2,000 hours, which ensures the life of more than 10 years.

CARTRIDGE

In order to maintain low error rate, the disk must be shielded to dusts, scratches and finger prints. The plastic cartridge is developed.

The sketch drawing and inside view of the cartridge are shown in Fig.8 amd Fig.9, respectively. Usually, the disk is fixed by clamp lever. When cartridge is loaded into thedisk drive, the shutter opens latterally to the motion of insertion, and at the same time , the disk clamp is released.

CONCLUSION

Large capacity and highly reliable DRAW type optical memory disk has been developed for HITACHI's file memory system, HITFILE 60.

By use of the combination of Te-Se-Pb alloy and cellulose nitrate layer for recording, raw error rate less than 10^{-6} is obtained for the capacity of 2.6 GB/disk (double sided). The disk thus developed has long life more than 10 years.

A new disk cartridge has also been developed, which improves reliability and offers easy handling of the disk.

REFERENCE

1) Tsunoda, Y. and Abe, S., "Optical File Memory for Computer Use" Technical digest of optical data storage.   1984, april

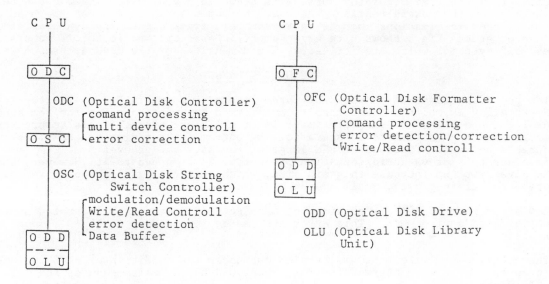

CPU

ODC

ODC (Optical Disk Controller)
  ⌈comand processing
  │multi device controll
  ⌊error correction

OSC

OSC (Optical Disk String
        Switch Controller)
  ⌈modulation/demodulation
  │Write/Read Controll
  │error detection
  ⌊Data Buffer

ODD
OLU

CPU

OFC

OFC (Optical Disk Formatter
        Controller)
  ⌈comand processing
  │error detection/correction
  ⌊Write/Read controll

ODD
OLU

ODD (Optical Disk Drive)

OLU (Optical Disk Library
        Unit)

(a) Large Scale System   (b) Small Scale System

Fig.1 Fundamental Construction of Optical Disk File System

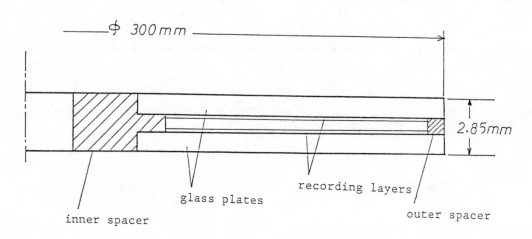

ϕ 300mm

2.85mm

inner spacer   glass plates   recording layers   outer spacer

Fig.2   Structure of optical disk

Fig.3   Outer view of optical disk

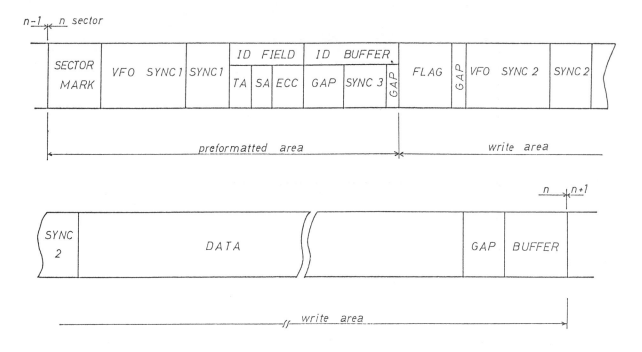

Fig.5   Sector format

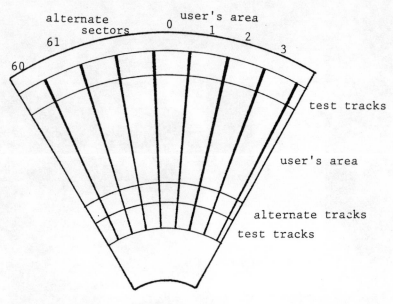

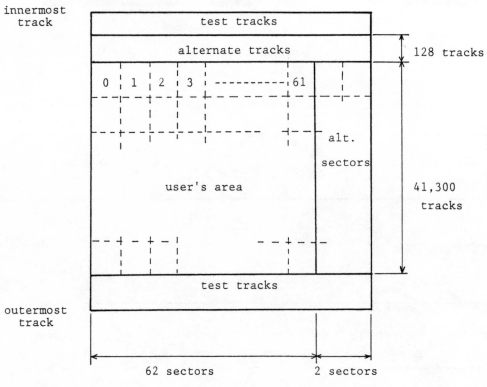

Fig.4   Data format on optical disk

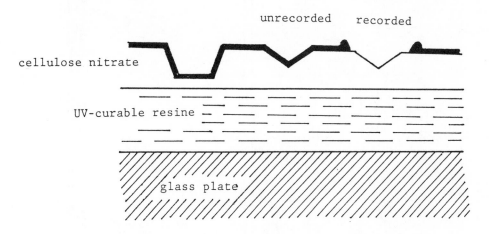

Fig.6   Structure of recording layer

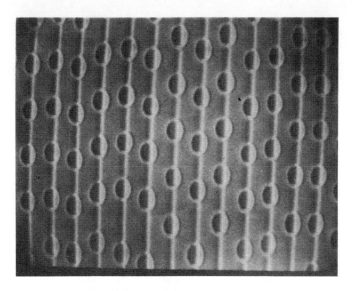

Fig.7   Pits shape on pregroove

(SEM image)

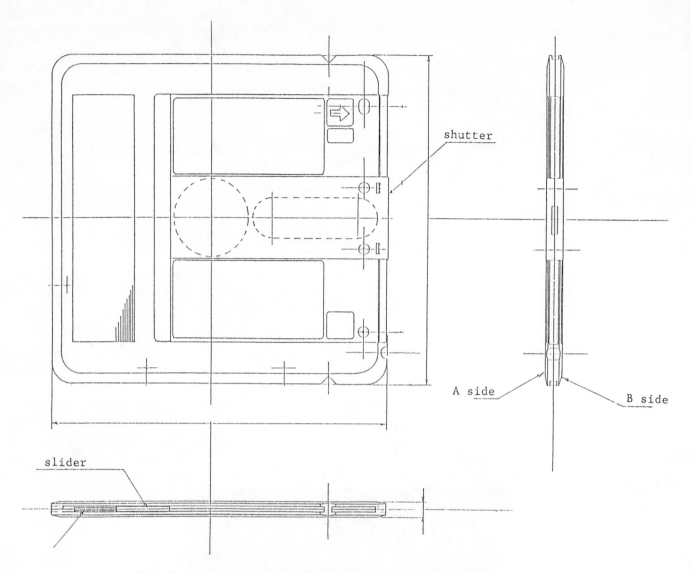

shutter

A side

B side

slider

Fig.8   Geometry of disk cartridge

Fig.9   Inside view of disk cartridge

Table 1   Fundamental specifications of optical disk

Capacity	2.6 GBytes/disk (double sided) By using OLU,83 GBytes can be stored (32 disks)
Reliability	bit error rate 10^{-12}
Data Transfer rate	1.5 MBytes/sec (Buffer memory in the controller) 0.44 MBytes/sec ODD itself
Average Access Time	0.25 sec/disk 6 sec (using OLU)

Table 2   Dimensions of the disk

Diamter	300 mm ϕ
Thickness	2.85 mm
Track density	16,000 TPI
Linear bit density	19,500 BPI
Thickness of glass plate	1.1 mm

Towards new information systems : Gigadisc

By Philippe Popoff and Jean Ledieu

Alcatel-Thomson Gigadisc
RN 186 la Boursidière 92350 Le Plessis Robinson (France)

Abstract of the lecture

Growth in knowledge has generated information in an ever-increasing volume and, simultaneously, has given rise to the need for its efficient distribution and long term pre-
servation. The new generation of storage media must be removable, possess a high recording
capacity and achieve substantially improved storage life over what magnetic media can
achieve for archival purposes.

Gigadisc meets all these requirements :

The Gigadisc digital optical disk of Thomson-CSF is a computer peripheral which uses
solid-state technology to store data. GIGADISC allows for the "irreversible" storage of
large volumes of digital data. Its 12" disk offers a capacity of one thousand megabytes
(one gigabyte) on each of the recording surfaces. Once written, the data can no longer be
altered. The disk is exchangeable and is protected within a cartridge which facilitates
manipulations of the disk. The disk together with the GIGADISC drive has been designed to
provide for an archivable life of over 10 years. The read-write unit includes a laser and
an optical head. It can access each block in the disk either at random or in sequence. The
SCSI interface allows for connecting GIGADISC to most computers. Its Controller can handle
up to 8 GIGADISC read/write units. Maintenance tests can be performed via a terminal
connected to the controller.

A high-level of performance is achieved by GIGADISC : the access time is of 10 ms within
a band of 40 tracks, 200 ms average beyond the current band, excluding 27 ms of average
latency. The GIGADISC read/write unit is supplied either in a rack-mounted chassis or as
a small stand-alone unit for installation apart on the top of a desk. This wide range of
facilities makes GIGADISC very attractive for running a great variety of applications in
the office automation, data processing and image storage/retrieval environment.

New potentialities

The growth in knowledge has generated information in an ever increasing volume and
simultaneously has given rise to the need for its efficient distribution and long term
storage.

These reproduction and distribution of information needs were facilitated by the disco-
very of printing, which has greatly reduced the volumes required for storing that information.

This was only with the advent of data processing, and more specifically with the magnetic recording techniques, that such volumes could be considerably reduced. The quality
of these techniques is continually progressing and allows for increased storage densities,
but is unable to satisfy the need of a dramatically growing market in the image storage
field, in particulars.

Hence, the requirement for a peripheral with a fast recording and retrieval performance
and for a high transmission speed has given rise to new techniques whereby images (scanners,
facsimile) and sound (vocoders, PCM) are digitized.

The synergy which is occuring between the office automation and the data (whether it is
made of data or documents) processing techniques has increased the need for a storage
mechanism capable of recording data, texts, digital images and speeches on the same media.

The media must be exchangeable and it must offer a large recording capacity as well a
long life-time archiving capability.

With these new facilities, the disadvantage which is inherent to the traditional magnetic media becomes obvious as their archival life is limited to two or three years, even
within a controlled environment.

Furthermore, the magnetic media are generally too expensive a tool for a lot of applications which one can foresee with a new media.

INFORMATION STORAGE WORLDWIDE ESTIMATES

Split into digitized and non digitized data

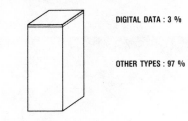

DIGITAL DATA : 3 %

OTHER TYPES : 97 %

1982 : 60,000 TERA-BYTES

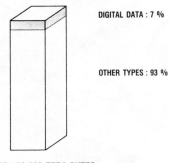

DIGITAL DATA : 7 %

OTHER TYPES : 93 %

1988 : 90,000 TERA-BYTES

Sources : DATA, IRD, REMY GENTON

Picture 1
Informations storage worldwide estimates

DIGITIZED INFORMATION WORLDWIDE ESTIMATES

Split into types of media

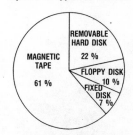

1982 : 1,800 stored tera-bytes

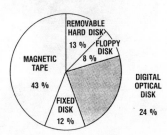

1988 : 6,500 stored tera-bytes

Sources : DATAQUEST, IRD

Picture 2
Digitized information worldwide estimates

GIGADISC, the new computer peripheral, is the best answer to these requirements.

Description of GIGADISC

The first product in the GIGADISC range has been worldwide launched by ALCATEL-THOMSON GIGADISC in November 1983.

The GD 1001 drive

The GD 1001 drive can be supplied in two models : a rack-mount model and a small size stand alone cabinet. The sizes of the rack-mount model are standard : 19"/4U(7X17.9X24.2 inch.) This model can be easily integrated within any compatible digital system. As much as the other model, it   can be connected to any computer, word-processor or to any system based upon a micro-processor. The drive is connected through the standard asynchronous SCSI (Small Computer System Interface) interface. This interface makes the connection of up to 7 GIGADISC controllers particularly easy to a computer. Each Controller is capable of handling up to 8 GIGADISC drives. Complex connections more than one computer, through the SCSI bus.

The lights and switches at the drive front panel provide for a set of friendly communications with the operator. These are control lights, two pushbuttons for loading and the write protect function, as well as a double-slot for loading and unloading the cartridge.

The internal structure of the drive has been designed as follows :

- an optical unit, which includes a solid state laser module and a photodetector ;

- an optical head, actuated for both radial and vertical positioning of the laser beam ("fine access") ;

- a linear motor, which focuses the optical head within the target track area ("coarse access") ;

- a rotational motor, which includes the disk seating and clamping device on its spindle ;

- a logic which controls the output power of the laser, the servo-mechanisms and the disk accesses.

Picture 3 : The GD 1001 drive

The GM 1001 disk

The 12" disk provides the user with an available capacity of one billion bytes in each of its recording surfaces. The disk is exchangeable and is protected within a cartridge which eases its manipulations.

The inner surface in the disk is coated with a bilayer which is made of a polymer and of a thin metallic films. While writing information, the laser beam evaporates the polymer layer, causing the thermal alteration ("bubbling") of the metallic film. Reading is achieved by the diffraction of the laser beam from the bubbles, which leaves these bubbles unaltered.

The disk is pregrooved and preformatted with a stamper. A continuous spiralled groove of 40,000 revolutions is radially divided into 25 sectors. The capacity of each sector is of 1,024 bytes.

This combined format of spiral and of sectorization allows for continuous read-write operations as well as for the random access to each sector.

Within less than 10 milliseconds, the optical head, without moving, can access 40 tracks i.e. one megabyte of data.

The error correction code feature diminishes the residual error rate from 10^{-5} down to a maximum of 10^{-12}.

The Mean Time between Failure (MTBF) of the GIGADISC drive exceedes 5,000 hours and its Mean Time to Repair (MTTR) is less than half an hour.

<u>Why the digital optical technology</u> ?

We previously saw the need for a low-cost, high recording capacity, easy to connect memory, which requirements GIGADISC answers at best.
New information systems can be now designed in combining digital optical disks, scanners, digital cameras, vocoders, high resolution screens, graphic printers. Such systems can operate in a stand-alone mode or connected to a Local Area Network and to a Wide Area Network.

These new information systems will integrate both magnetic and optical peripherals, each type of peripheral offering different specifications and meeting different purposes in terms of capacity, functionnality and cost :

The magnetic technology allows for the permanent recording of data within a period of 2 or 3 years, as a maximum term ;

The optical technology enables to record larger amounts of data, be preserved without modification for much longer periods of time and at a much lower cost.

Hence, these two technologies are complementary to each other. The diagram herunder shows the characteristics of the digital optical technology, compared with the magnetic technology.

TAPE AND DISK DRIVES

COST PER MEGABYTE, ON LINE

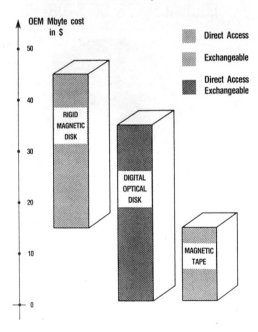

<u>Picture 4</u> : Tape and disk drives comparizon

The optical technology market segment is aimed at archiving a large amount of data at low cost and therefore the paper replacement business is the true market area for this technology.

<u>The applications of the technology</u>

The memory peripherals can be split over the long term into four major technologies :

- the current magnetic technology ;

- the rise in 1985 of a new technology based upon the vertical magnetism ;

- the actual rise of the non reversible optical technology ;

- the emergence in 1986-1987 of the reversible optical technology.

The major typical applications for which GIGADISC is the most appropriate answer are known as :

- storage of digitized data of all types where the real time access facility to each record is required ;

- archives which require to be preserved over more (or even much more) than 10 years ;

- chronological information systems which, by definition, do not need a high level of updating.

The following diagram shows the comparisons in storage capacity between the traditionnal media and GIGADISC.

EQUIVALENCES OF STORAGE

1 DOUBLE-SIDED DISK
GIGADISC   **=**
GM 1001 / 121
(16,400,000,000 bits)

- **DOCUMENTS :**
  — 40,000 sheets   (A4)
  — 400 microforms (A6)

- **COMPUTERIZED DATA :**
  — 655,000 sheets   (A4)
  — 3,200 microforms (A6)
  — 20 magnetic tapes (100 MBytes)
  — 10 removable cartridges (200 MBytes)

- **AUDIO :**
  — 1,800 hours (LPC 2,400 bits/sec.)
  — 70 hours (PCM 64 Kbits/sec.)
  — 3 hours (HI-FI CD compatible)

<u>Picture 5</u> : Equivalences of storage

Among several potential applications of the digital optical disk, archive systems may be predicted as the most promising in all areas.

- In the office automation sector area, GIGADISC is the cornerstone in dedicated information systems where digitized data (documents, images, voice, sound) or coded data (texts, computer files) are to be accessed. Such systems can store, retrieve, process, print, replicate and transmit the information through a Local Area Network or through a Wide Area Network.

- In the data processing sector area, GIGADISC offers, at the lowest cost, the smallest volume for such applications as : data bases, large volume log and transaction files, printing files, back-up files ...

  The non reversible digital optical disk can help the problem of the software copyrights, by the non-erasibility of the original license number.

- In the documentation sector area, where books, engineering drawings, films, medical files, etc. are difficult to handle, GIGADISC is particulary efficient and convenient.

- In the military sector area, GIGADISC is well suited in the documentation, maintenance and training services application fields. It guarantees a high level of data reliability (10^{-12} of remaining errors after correction), a very compact size, and furthermore the disc is not sensitive to electromagnetic fields.

This wide variety of applications are extremely good applicant to the digital optical disk, which could then become a worldwide standard media for archiving data.

The digital optical disk market

In the long run the main characteristics of the storage peripheral market, in the EDP and office automation sector areas, are :

- an average annual growth of 20 %, starting at $ 15 billion in 1984 and moving up to $ 31 billion in 1988.

- the digital optical technology is expected to grasp 7.4 % of this market, i.e. $ 2.3 billion in 1988, with an annual average growth in excess of 40 %.

THE DATA STORAGE PERIPHERAL MARKET

TECHNOLOGY BREAKDOWN

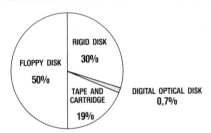

1984 : 15 billion Dollars

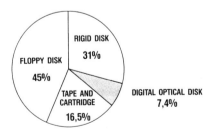

1988 : 31 billion Dollars

Origin : ADL; DATAQUEST; MACKINTOSH

Picture 6 : The data storage peripheral market technology breakdown

THE DATA STORAGE PERIPHERAL MARKET

GEOGRAPHICAL SPLIT

1. ALL TECHNOLOGIES

2. OPTICAL TECHNOLOGY ONLY

Picture 7 : The data storage peripheral market geographical split

The digital optical market is shared into three segments, which respectively are made of the drive (50%), of the controller (10%) and of the media (disk) (40%).

Depending on the system performance factor, one can consider what follows :

- A high-end market (14" disk, more particularly designed for the main frame computers);

- A mid-range market (12" or 8" disks) ;

- A low-end market (5" or 3.5" disks) ;

- A market for mass-storages (more than 50 gigabytes on-line).

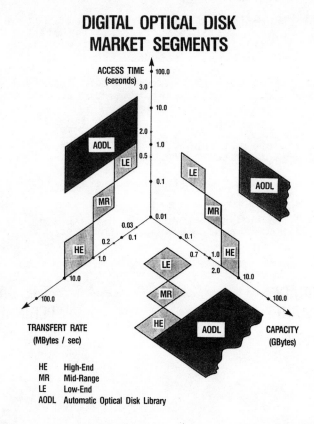

Picture 8 : Digital optical disk market segments

The low-end market segment is expected to become the largest in volume from 1988 onwards due to its use in the office automation sector and with personal computers.

Conclusion

GIGADISC is a new low-cost peripheral which is particularly suited for being integrated within new system architectures.

Manufacturers, designers, system integrators may today develop a variety of products and systems. The GIGADISC GD 1001 is the first product which has been announced by the ALCATEL-THOMSON GIGADISC Company, which stands as a startpoint to a new storage golden age.

The Role of Standards in the Emerging Optical Digital Data Disk
Storage Systems Market

Ross C. Bainbridge

Institute for Computer Sciences and Technology, National Bureau of Standards
Washington, DC 20234

The Institute for Computer Sciences and Technology at the National Bureau of Standards is pleased to cooperate with the International Society for Optical Engineering and to join with the other distinguished organizations in cosponsoring this conference on applications of optical digital data disk storage systems.

I will discuss standards for this emerging technology within the context of NBS' activities to promote the more effective use of information technology.

The National Bureau of Standards provides technical support to both government and industry in the beneficial and effective uses of information technology. The development of standards for computers and networks is just one part of our overall responsibilities. We are a source of technical advice and assistance to users of computers and to the organizations that develop policies for the technology and its uses. The technical foundation for our work is our laboratory-based research to investigate new applications of technology and to develop test methods and measurement techniques.

In all of our programs we are keenly aware of the many organizations that are affected by computers. We cooperate with many different groups to transfer technology, to share information about what we are doing, and to learn about what others are doing.

The technical areas that we are addressing are all aspects of information exchange-- between parts of a system, between humans and systems, and between humans and other humans concerning expectations, benefits and risks in using computers and networks. Standards are often the discipline that we adopt to achieve this interchange. Standards are key to the consumer's interest. They are important in assuring the quality of goods and services and the interchangeability of parts. But perhaps most important of all, standards are essential to the communication of ideas and technical information.

We believe that it is important to work directly with industry and with users to determine their requirements for technical support and for standards. We do this through our contacts with computer users and with manufacturers who want a wide market for their products. We are frequently consulted by Congress, Federal government agencies, State and local governments, and by private sector users for technical advice on computers. Our exchanges with users and manufacturers through conferences such as this one, and through workshops, and seminars that we sponsor, are ways that we learn about their requirements and foster the exchange of information on new technology, its applications, risks, and benefits.

When there is a clear need for standards, we work through the voluntary industry processes to help build consensus and advance the development of the needed standards. The issue of timing is key to the development of useful standards. Standards developed too early in a technology's life cycle may be barriers to further innovation. Standards developed too late are often "de facto" standards that are based on a specific product. However, standards developed at the right time can help to advance innovation and to aggregate the market for new technical products.

In working with voluntary industry standards organizations, NBS's goal is to encourage the development of off-the-shelf commercial products. This helps to broaden the base of potential users of standard products. NBS helps to write the technical specifications. But writing standards does not assure that useful products will result. Standards must be implemented properly in order to be useful. NBS draws upon its laboratory research to develop the test methods and measurements techniques to assure correct implementation and performance. This testing is often done cooperatively with users and manufacturers.

NBS has been actively involved in the development of standards and test methods for computer storage media since the late 1960's. We contributed to the development of voluntary industry standards for magnetic tape, tape cassettes, and cartridges. We are currently active members of standards committees of the American National Standards Institute, the International Organization for Standardization and the Association for Information and Image Management.

Our research efforts include calibration services, the development of measurement methods and standard reference materials for computer storage media, and investigation of the properties and characteristics of the media. NBS developed standard reference materials for half inch wide digital magnetic computer tape, digital magnetic tape cassettes, digital magnetic tape cartridges, magnetic flexible disk cartridges, and OCR-B centerline drawings. We are the sole worldwide source for the tape reference materials. The use of standard reference materials has noticeably improved the quality of storage media products and improved the reliability of data exchange for the user.

The emergence of optical digital data disk technology opens new opportunities for users to automate processes that could not be easily automated previously and to take advantage of the expected economy and improved storage capacity of this new media over existing storage media. We have started an effort to transfer information about the technology's potential advantages and limitations and to discuss user and vendor needs for technical support. Last year we cosponsored a workshop with the National Security Agency to discuss applications of OD^3 technology and potential requirements for guidance, and standards. The proceedings of the workshop were published earlier this year and distributed broadly to both users and vendors. We also issued a comprehensive bibliography of references to literature on optical storage technology to aid researchers in this area.

We plan to establish a council of Federal government representatives to exchange information on the technology and its applications. This group will hold its first meeting in September 1984. However, before users can rely on OD^3 technology for their long term storage media applications, there are many questions to be answered. NBS will use its research capabilities to investigate some of these questions--the long term stability of the disks and their susceptibility to data losses.

We will continue to work with both manufacturers and users to evaluate the technology, to provide technical support, and to develop of standards and test methods when they are timely and appropriate.

APPLICATIONS OF OPTICAL DIGITAL DATA DISK STORAGE SYSTEMS

Volume 490

Session 2

Applications

Chairman
Ann U. Kerr
DARPA, USA

Application of the digital optical storage technology
to the management and processing of satellite data

Patrice Boursier

Laboratoire de Recherche en Informatique, University of Paris-Sud
91405 ORSAY Cedex, FRANCE

Abstract

During the last fifteen years, earth observation satellites like LANDSAT or SEASAT have been revealed as tools of prime interest for the acquisition of a specific type of information said to be remotely-sensed. Many fields have taken advantage of the analysis and processing of this information. This is the case for such applications as land-use planning, water resource management or pollution prevention. Most of the time however, satellite data do not bring as much as they could. The benefit resulting from their collection is indeed limited, due to the huge amount of information they represent, and the difficulty or even the impossibility to process all of them. In most cases, they are stored on magnetic tapes either in original or in a pre-processed form, and the analysis of the frames is done on an individual base, except for large installations that permit to manage large scale studies (e.g. weather forecasting). The use of the digital optical disk will eliminate this limitation by allowing the on-line storage and processing of vast amounts of remotely-sensed data, thus bringing them an extra-value. This is what we want to show within a few examples.

1. Introduction

It is a long way since the launching in 1972 of LANDSAT-1, the first one of a series of earth observation satellites. It is only twelve years since that time, but during that short period, a lot of improvements have been brought to the management and processing of the data thus collected as well as to the characteristics of the sensors.

Many fields have been concerned and have taken advantage of this new and still evolving source of information. Weather forecasting, natural resources inventories, protection of the environment have been among the first to benefit of the use of satellite data.

However, this optimism must be tempered by the fact that remotely-sensed data represent a huge amount of information to be handled and processed.

The emergence of such very large mass storage as the digital optical disk will certainly change many habits and open new horizons, but one should not think that all the problems are solved, because it is not the case. Important progress can yet be foreseen by the joint use of this new techonology and other hardware components that have improved a lot during the last years. We particularly think to the increasing power and capacity of processing units and central memories, as well as to the astounding possibilities of graphical devices such as colour graphic CRT's and laser printers.

Important advances are also to come from the side of software, and especially from the field of image-oriented data base management systems.

Typical applications, as well as problems and limitations in the use of remotely-sensed data are approached in section 2. The contribution of data base management systems is discussed in section 3. Finally, section 4 tries to outline the future of image information systems. It shows in particular what may be the impact of the digital optical technology in the development and use of such systems.

2. The use of satellite data: applications and limitations

2.1. Typical applications

The number of applications handling satellite data is always increasing, and the aim of this paper is not to give an extensive description. We think that it is yet important to make a brief survey, so as to situate the problem. It may seem hasardous to make a classification of such applications, but for convenience sake we will distinguish the "air-related" applications, the "sea-related" and "earth-related" ones. We will also make the difference between the short-range and the medium or long-range applications, and between those which are geographically restricted and those which are extended. These classifications will be useful in the following to characterize and consider the future of the use of remotely-sensed data.

a) "Air-related" applications.
Weather forecasting (short-range, extended area) and all kinds of climatological studies (medium/long-range, restricted area) may be placed here. Data are obtained via the METEOSAT and TIROS series satellites.

b) "Sea-related" applications.
We include in this class the help for fishery (short-range, restricted) and such oceanographic studies as marine geology, living marine resources or ocean contamination (medium/long-range, extended). The SEASAT, NIMBUS and METEOSAT satellites are used to collect these data.[7,9]

c) "Earth-related" applications.
They are the most numerous and deal with data transmitted by the LANDSAT series satellites. Covered fields are very diverse since they include the help for agriculture, geology, hydrology, natural resources protection and inventories, or land-use planning.[1,8,15] Most of the time, these are medium or long-range applications, limited to a small geographic area.

2.2. Problems and limitations

They have two origins: the nature of geographical data and the characteristics of the sensors.

a) Nature of geographical data.
Geographical data are a very specific kind of information that require adequate storage structures and processing methods.[3,6,13,17] Moreover, the quantities of information to be handled and processed are generally very large.

b) Characteristics of the sensors.
Problems inherent in satellites are related to the reliability of transmitted data ("noise" due to natural perturbations such as clouds, or due to the sensors), and to the huge amount of information they constitute. This last difficulty is increased by the repetitivity of data collection, and things are not going to turn out better since the number of satellites is still increasing and their resolution goes thinner and thinner. Let us give an example with the data transmitted by LANDSAT: with a 80 meters resolution (Multi-Spectral Scanner), 8 Mbytes of storage were necessary to store only one single-channel LANDSAT image! Now, the resolution of the LANDSAT-D Thematic Mapper is 30 meters and the soon-to-be-launched european SPOT satellite will offer a 10 meters resolution!

All this information cannot be indefinitely and fully archived, and many studies cannot be done because they are related to a very specific and gegraphically restricted phenomenon that is noticed too late. The problem is that forecasting and planning is needed when it is almost impossible to do so.

3. Data Base Systems for the management and processing of satellite data

The interest in the use of a DBMS for handling satellite data is not different from the classical DBMS interest.[16] But here again, the nature of the processed information bridles the development of such systems. It is yet symptomatic to notice that only a few prototypes have just been designed and implemented.[2,10,11,12] The problems they may help to solve are in any case limited both qualitatively (observed parameters), and quantitatively (covered area).

4. The future of satellite data management: image information systems ?

4.1. Possible improvements

DBMS rapidly evolve, and the number and variety of interested fields is continually increasing, but it is still a long way to the use of real image information systems. Concerning their use for the management and processing of satellite data, valuable improvements and/or rapid progress may be anticipated in the following areas:

a) Data storage hardware.
The digital optical technology can bring a lot in that domain.

b) Data processing hardware.
Specialized parallel architectures for image processing are under development and rapidly progress.

c) Hardware for the display and printing of images.
Colour graphic CRT's now allow to display images containing more than 1 Million of pixels, with a great number of different colours. Top-level laser printers can produce documents that stand comparison with classical "hand-made" ones.

d) Image management software.
Progress has not been so fast in that field. Packages and prototypes have been designed and sometimes implemented, but they are in most cases experimental, or their use is limited to specific kind of applications.[4,5,14] We yet believe that in the near future, real friendly and performing systems will appear. In that way, a real "image data model" has still to be defined.

e) Image processing software.
The major limitation stands here, and a lot of work is to be done to be able to really operate on the information that is "contained" in an image.

4.2. Contribution of the digital optical technology

The arrival of digital optical storage is important for several reasons:
. the storage capacity is now expressed in terms of giga-bytes instead of mega-bytes,
. the access time to information may be compared to that of magnetic disks,
. the cost for storing very large quantities of data is relatively low,
. the medium is reliable and robust, and his life is expressed in years.

It becomes then possible to archive and/or have "on-line" an impressive quantity of information, thus opening the way to new applications and allowing to extend the range of already existing ones. This will be the case with satellite data management, since it will be possible to enlarge the areas under consideration, even if restrictions still subsist concerning the processing of these data. It will also be possible to lay out such large scale studies as land-use planning that require to take into account a large number of parameters.
Besides, it will be possible to archive more information, and for a longer time, thus allowing late medium and long-range studies to be made.
But it is also worth-noting that specific data structures will have to be designed so as to manage and process image-like information with digital optical disks.

5. Conclusion

We wanted to show here what can and will probably be the impact of the digital optical technology in the management and processing of data obtained via satellites.
We have briefly surveyed the applications of such data, and evoked some problems and limitations, both due to the sensors and to the specific nature of the information taken into account.
The digital optical disk surely brings elements to solve these problems, but it does not bring "the" solution. Then, other important factors have to be considered. We have in mind the hardware evolution as well as the development of real image information systems that will allow to manage and process image-like information.

Acknowledgements

The author wishes to thank Catherine Mering, from the Remote-sensing Laboratory at the Office de la Recherche Scientifique et Technique Outre-Mer (ORSTOM) for helpful discussions and documentation.

References

1. Bernstein R., "Data base requirements for remote sensing and image processing applications", Lecture Notes in Computer Science, Vol. 81, pp. 319-345, 1980.

2. Billingsley F.C., "Data base systems for remote-sensing", Lecture Notes in Computer Science, Vol. 81, pp. 299-318, 1980.

3. Boursier P., Scholl M., "Performance analysis of compaction techniques for map representation in geographic data bases", Computers and Graphics, Vol. 6, No. 2, pp. 73-81, 1982.

4. Boursier P. et.al., "SYLVAIN: a microcomputer-based and relational-like cartographic database management system, designed as an integrated tool for land-use planning", IEEE Conf. on Systems, Man and Cybernetics, Bombay and New-Delhi (India), 30 Dec. 1983 - 7 Jan. 1984.

5. Chang S.K., Kunii T.L., "Pictorial database systems", IEEE Computer, Vol. 14, No. 11, pp. 13-21, Nov. 1981.

6. Chock M., Cardenas A.F., Klinger A., "Manipulating data structures in pictorial information systems", IEEE Computer, Vol. 14, No. 11, pp. 43-50, Nov. 1981.

7. Domain F., "Le satellite METEOSAT et l'oceanographie", Cahiers de l'ORSTOM-Teledetection, No. 3, Paris, 1979.

8. Guillot B., "Les satellites de l'environnement", Cahiers de L'ORSTOM- Teledetection, No. 5, Paris, 1981.

9. Hibbs A.R., Wilson W.S., "Satellites map the oceans", IEEE Spectrum, pp. 46-53, Oct. 1983.

10. Kondo T. et.al, "A map-guided image database system for remotely-sensed data", IEEE Conf. on Pattern Recognition and Image Processing, Las Vegas (USA), pp. 333-338, June 1982.

11. Lohman G.M. et.al., "Remotely-sensed geophysical databases: experience and implications for generalized DBMS", ACM Sigmod Record, Vol. 13, No. 4, San Jose (USA), pp. 146-160, May 1983.

12. Nagata M., "A relational image database system for remote sensing", IEEE Conf. on Pattern Recognition and Image Processing, pp. 491-495, 1981.

13. Nagy G., Wagle S., "Geographic data processing", Computing Surveys, Vol. 11, No. 2, pp. 139-181, June 1979.

14. Sloan K.R., Lippman A., "Data bases of/about/with images", IEEE Conf. on Pattern Recognition and Image Processing, Las Vegas (USA), pp. 441-446, June 1982.

15. Teicholz E., "Processing satellite data", Datamation, Vol. 24, pp. 117-132, June 1978.

16. Ullman J.D., "Principles of database systems", Computer Science Press, New-York, 1980.

17. Zobrist A.L., Nagy G., "Pictorial information processing of LANDSAT data for geographic analysis", IEEE Computer, Vol. 14, No. 11, pp. 34-41, Nov. 1981.

Large databases in a meteorological environment

Peter Gray, Dick Dixon, Geerd-R. Hoffmann

European Centre for Medium Range Weather Forecasts
Shinfield Park, Reading, Berkshire RG2 9AX, England

Abstract

The European Centre for Medium Range Weather Forecasts (ECMWF) has embarked upon a project to provide a back-end file server subsystem to a variety of worker machines ranging from a VAX 11-750 to a CRAY X-MP. The project commenced in mid-1982 with a procurement phase, which lasted almost 12 months and is presently in the implementation phase. It is expected that a significant proportion of the project will be completed during 1984.

The file server is based upon IBM architecture and will be connected to the worker machines using a 50 million bits per second proprietary local area network supplied by Control Data Corporation.

The file server will support a large meteorological database which currently contains around 2×10^{13} bits of information and which is growing at a rate of 3×10^{12} bits per year. A variety of magnetic storage devices (Disk Mass Storage System and Tape) will provide a hierarchical storage system which is well placed to be able to use new devices as they emerge.

The paper will describe in detail the application, and the hardware and software necessary to realise the solution adopted at ECMWF. It will also examine its requirements for new storage devices, with particular emphasis on Optical Digital Data Disk Technology.

About ECMWF

The European Centre for Medium Range Weather Forecasts (ECMWF), located at Shinfield Park, Reading, England, is an inter-governmental organisation established by an International Convention in 1973 to carry out the development of numerical methods for medium-range weather forecasting (i.e. currently for up to ten days ahead), and to prepare and distribute regular medium range weather forecasts to its seventeen European Member States.

In order to adequately carry out this role, the Centre engages in continuing scientific and technical research, and also maintains a considerable data processing and storage operation. This is presently supported by a CRAY X-MP and two Control Data Cyber 800-series processors, plus numerous smaller machines.

The data stored by the Centre currently occupies approximately 25000 reels of industry-standard (2400 ft. x ½ inch) magnetic tape. This figure is growing by about 4000 reels per annum: one reason why the Centre determined in 1982 that an alternative approach must be found.

The Centre's strategy in attacking the problem of data storage has been to transfer the bulk of medium and long lifetime data to a small mainframe computer dedicated to the data storage and retrieval function. It is hoped that by this means the administration of the storage media can be made more efficient and the duplication of stored data reduced (so helping with the data volume problems), and that adaptation to new requirements and newly available technologies can be made more easily (thus offering long-term continuity).

ECMWF applications and requirements

Applications

The data to be stored and accessed can be split into 4 general types:

(i)   Reports. The Centre gathers meteorological observations continuously from the whole globe. At present around 10 million bytes per day are collected, but this quantity will increase dramatically when newer observing systems become available. The data is checked and obviously erroneous data removed, and is then stored indefinitely. No subsequent updating of the data is performed.

Occasionally, special observing experiments are undertaken by the world meteorological community which result in a period of intense observational activity and create substantial databases containing observations plus a few fields. Such data is also stored indefinitely, but may occasionally be updated or reprocessed.

(ii)   Operational fields. A field is a matrix representing a meteorological variable at some atmospheric level, at a given stage in the forecast. At present around 200 million bytes of such fields are produced daily by the Centre's production forecasts, but plans for improved forecasting systems could increase this number by a factor of 8. Data from an operational forecast is not updated, and is normally kept for a period of at least 5 years.

(iii)   Experimental data. Experimental data is normally in the form of fields created by forecast systems under development, and around one thousand million bytes per day are currently created. This data is usually kept for a period of six months to two years.

(iv)   General data. This category includes user programs, computer service statistics etc., which do not fit into any of the above classifications. Data volumes are of the order of 100 million bytes per day, and retention periods vary between 1 day and infinite. These files will often be regularly updated.

Data usage

It is almost impossible to predict the demands which will be made on the new centralised filing system from the pattern of use of the existing archiving system. This is because the present system handles only reports and operational fields, and because it handles the data in a very different way from that in which it will be handled in the future. Thus the estimates given must be somewhat conjectural.

Nevertheless, it is anticipated that up to 3 thousand million (3×10^9) bytes of new data must be stored each day, which will generally be retained for between 180 days and 5 years. A small amount of data, (perhaps 100 million bytes per day), will be retained indefinitely.

It is even more difficult to estimate retrieval rates, but it is expected that around 1 thousand million bytes will be retrieved daily. It is impossible to accurately predict the probability of access to any particular file.

It is expected that the average file size will be of the order of 1 million bytes. File sizes will range from a minimum of 2 thousand bytes to a maximum of 100 million bytes: both of these limits are in practice set by the filing system used at the Centre.

The strategy adopted

The commercial availability of high-speed computer interconnection systems (known as Local Area Networks, or LANs) made possible the contemplation of a separate, dedicated computer system devoted exclusively to data storage and retrieval. A period of intensive study led to the conclusion that this approach would offer considerable advantages in terms of efficiency, resilience and adaptability, and an ITT (Invitation to Tender equivalent to an Operational Requirement or Request For Proposals) was accordingly prepared in October 1982.

Evaluation of the responses to the ITT led to the decision that the system should be based on hardware and software systems from several sources. At the present time, a subset of the hardware is installed, and software development and integration are proceeding.

A local area network of 50 million bits per second capacity, supplied by Control Data Corporation, provides the means for interconnecting the Centre's mainframe computers with the new file handler. The network will connect 3 CDC Cyber 800 series mainframes (one used as a gateway to the CRAY X-MP), a VAX 11-750 and the IBM 4341 file handling processor.

The file handling system hardware consists of an IBM 4341 processor with 3380 discs, a 3850 cartridge mass storage system and conventional 6250 bpi magnetic tape units. A significant factor in reaching this decision was the expectation that vendors of new data storage devices would provide early support for IBM systems in order to attack the largest possible market. (The recent announcement of IBM's 3480 tape subsystem and Storage Tek's optical disc demonstrate the validity of this approach).

Media characteristics are discussed later in this paper, but it should be noted that no immediate change is expected in the Centre's existing strategy of using conventional half-inch magnetic tape as the only long-term, large-scale storage medium. The critical problem with the rate of growth of the Centre's tape library is, for the moment, being tackled purely by the improved media and data management which the new system will permit.

The basic software for the file handling system is the "Central Filing System"(CFS) from Los Alamos National Laboratories in the USA, which is in use at a number of US Government sites. The characteristics of CFS are:

(i)   It is able to handle very large amounts of data (currently more than 10^{13} bits at Los Alamos);

(ii)   it is able to store many files (currently 1.25 million at Los Alamos);

(iii)   it interfaces to worker machines via a local area network;

(iv)   it is able to efficiently use a variety of data storage hardware. Currently IBM 3330, 3350 and 3380 discs are used, with both online and offline 3850 MSS cartridges. Support for magnetic tape is being added and experience shows that support for new devices can be added fairly easily.

From the point of view of possible use of optical data storage technology, this approach has considerable advantages, summarised as follows:

1.   IBM processor provides the most popular hardware interface.

2.   Indirect attachment to data processing complex minimises impact of new devices on users of the system.

3.   Sophisticated management software permits use of a mix of existing and new technology.

4.   Centralisation of file storage function permits economic use of large-scale data storage devices which would otherwise be under-used.

The next section reviews the characteristics and shortcomings of existing data storage devices in the Centre's application.

Existing hardware solutions

At the time of procurement (late 1982), the following commercial devices for data storage hardware were identified:

(i)   <u>Magnetic disc</u>. In particular, the latest generation thin film head technology devices such as the IBM 3380 offered attractive characteristics. For the Centre's application, they had the advantage of being fast, reliable and well-supported, but suffered from a high cost per bit, and from large floorspace and environmental requirements. The location of large amounts of data under a single accessor mechanism, which is increasingly imposed by the latest generation devices, is not a real problem for a centralised file handling system, whereas for a general-purpose mainframe computer it may well be so.

(ii)   <u>Magnetic cartridge storage devices</u>. The IBM 3850 or the more recent Masstor M860 were available at the time. These have lower cost and smaller environmental requirements to offer in exchange for much slower access times, but were still too expensive, even in media cost alone, to contemplate their use for the Centre's entire data storage requirements.

(iii)   <u>Half inch magnetic tapes</u>. The media cost of conventional tape would be acceptable for the Centre's data volumes, and the long term availability of software and equipment to read them is likely to be excellent (an important factor for long-term data archives). However, the routine imposition of long reel-mount delays could not be tolerated (although for special requests and infrequently used data it could be so), and the manual administration of a large tape library would impose an additional manpower cost.

(iv)   Automatic Tape Library systems. The only system commercially marketed was not offered in any response to the Centre's ITT. In any event, the storage limitation of 9000 tapes with the existing device would have been unacceptable, and the absence of any similar installation in the UK would undoubtedly have created considerable support problems.

Evidently, none of the above devices could alone offer an adequate solution. The classic solution to this dilemma is to employ a mix of storage devices, with software to manage them in a hierarchic fashion; and indeed the initial ECMWF system is based on a magnetic disc - automatic cartridge library - magnetic tape hierarchy.

However, even this system, the best that could be obtained, suffers from the following disadvantages:

.   A complex software system is needed to manage the hierarchy of devices.

.   The ratio of file access times is approximately 1:1000:20000 for files resident on disc, cartridge and tape respectively.

.   The use of 6250 bpi tape suffers from the usual problems of media maintenance and manual handling.

The CFS software system has, in the past, been found to be adaptable to new storage devices, and it is clear that digital optical disc technology could readily be utilised. CFS has three characteristics which overcome some traditional problems with digital optical systems:

(i)   The media holding CFS files contain no information which must be updated when a file is used or deleted. All administrative information relating to CFS files is stored in a centralised directory, normally a VSAM file on 3380 disc.

(ii)   CFS never updates files in place. CFS files are always updated by creating a completely new version, then updating the directory to refer to the new version. Similarly, CFS file deletion does not require any access to the data storage medium, and takes place purely in the directory.

(iii)   For a "worker machine" (i.e. one of the general-purpose mainframes), all access to CFS-managed data is via the local area network and associated software. Thus, unique media characteristics can be hidden from the users of the system, and device compatibility is not an issue.

Adding optical discs to the Centre's system

As should be apparent from the foregoing discussion, the Centre's environment offers a uniquely favourable opportunity for the incorporation' of optical discs in a large scale data storage operation, partly because the separate file handling system effectively "decouples" the application from the characteristics of the storage media.

From the Centre's point of view, optical disc would provide the following advantages if it were available as an extra element in the storage hierarchy:

i)   Potentially low media cost (deriving largely from expected high media capacity): if media cost were competitive with magnetic tape, an alternative strategy for dealing with unbounded growth in stored data volume would be available.

ii)   High media density: the physical bulk of data stored on optical disc would be expected to be at least an order of magnitude less than that of the same data on magnetic tape.

iii)   Rapid access: the data transfer rates and accessor positioning delays would reasonably be expected to be comparable with conventional magnetic discs, giving a speed of access to online data which would not be matched by any tape-based device. Nevertheless, access to offline data would still be dominated by manual volume retrieval and mount/dismount operations (which might be adversely affected if there were, for example, long spin up/spin down delays). For very large data storage operations such as the Centre's, therefore, use of optical disc media instead of magnetic tape might not offer a significant improvement in access delays unless an automatic volume mount system ('juke box') were available.

Although the first two items are important considerations, any advantage in these areas alone would need to be of very large magnitude to justify a move away from conventional magnetic media. In particular, a situation where media cost and mount delays were comparable with magnetic tape while access speeds and device cost were comparable with magnetic disc would appear to offer few convincing arguments in favour of optical technology. Hence, it appears that the automatic volume mount facility would have to be made available before optical storage became attractive for the Centre (of at least 10^{12} bytes capacity, and preferably ten times that).

Some further important considerations which would be crucial to decisions on use of optical storage:

i)   Media compatibility: total data compatibility between all devices at one installation must be guaranteed.

ii)   Data error rates must be equal to or better than the best currently offered by rotating magnetic media.

iii) Short-term reliability must be at least comparable to that of currently available disc drives.

iv)   Long-term media reliability must be such that no additional hard errors would be expected following a five-year storage period.

v)   Software support must be available offering IBM/MVS facilities at least up to a level comparable with BSAM.

vi)   Optical discs employ highly advanced technology: it would be disastrous if day-to-day maintenance were such a highly-skilled operation that availability of personnel became a problem.

Two further points bearing on the suitability of optical storage for the Centre's application are: firstly, that the famous "write once" characteristic is probably not a problem (partly because of the expected low update rate of the Centre's data and partly because of the management characteristics of the CFS file handling software), and secondly: that before the Centre could commit to changing the basis of its long-term data archives from magnetic tape to any other medium, it would need to be assured, not only of the reliability of the media and the devices in such an application, but also that the supplier and the necessary support services would continue to exist for an extended time, and that the particular device format would not shortly become obsolete and cease to be supported. The cost of migrating a 2×10^{12} byte data base to a different medium is not easily met.

Conclusion

In considering optical technology, the Centre is looking for a reliable system on which it can safely store its archives and which has a large total online capacity in order to reduce the costs of handling media volumes. In addition, although its access time need not be as fast as magnetic disc, it must be substantially faster than magnetic tape, at a media cost comparable with a magnetic tape library system.

The substantial cost/performance gap between conventional magnetic disc and tape, which is not adequately filled by existing magnetic cartridge devices, means that a ready market exists for the optical disc if the cost, performance and capacity parameters can be got right. The unique characteristics of the device are most easily accommodated in an environment such as the Centre's, where files are only indirectly presented to users of the system.

Nevertheless, the Centre has special requirements in terms of long-term maintainability and very large total data volume. It remains to be seen whether a commercial device matching all of these parameters will become available in the near future.

A DOCUMENT STORAGE APPLICATION : THE SARDE PROJECT

GILLES CHAUVIN, MICHEL PICARD

CENTRE NATIONAL D'ETUDES DES TELECOMMUNICATIONS
38-40, rue du Général Leclerc
92131 ISSY-LES-MOULINEAUX FRANCE

ABSTRACT

SARDE project aims to replace a 5 million pages technical documentation, partly reproduced in 2000 sites, by a fully electronic storage, retrieval and display system. This system architecture is as follow : 1/ the acquisition of documents (their format is A4 to AO) is done by scanners ; the documents are then processed ans compressed, 2) the document storage uses several THOMSON GIGADISC, supported by a dedicated architecture, able to provide high efficiency, great reliability and easy exploitation ; juke-boxes can be used ; 3) the documents are accessed by a classical data base ; 4) the documents are sent to remote users through 64 Kbits links ; 5) remote users are provided with a workstation, consisting of a high definition screen (4 million pixels, 19 " size), a powerful microcomputer, a small image printer, local disk and network interface.

A prototype system is to be built for 1985, and experimented in real context with end users.

1. THE SARDE PROJECT : OBJECTIVES

The French Telecommunications Administration shows a great interest in Electronic Documentation Systems. A first one was PAPYRUS, realized by Integrated Automation and MC2, for a library application at the C.N.E.T It was one of the first utilization of the THOMSON GIGADISC optical digital data disk. More details on this system can be found in (1).

The C.N.E.T which is the National Research Center of French Telecommunications has a project called SARDE described in this paper.

The long term goal of the project is to change totally the access, storage, circulation and management of the technical documentation of the French telecommunications network. The basic ideas are as follow :

- technical documentation is of very great importance for a good quality of exploitation and maintenance of our telecommunication exchange centers, - but costs and working load must be kept moderate,

- traditionnal means, including the use of microforms for storage and data bases for retrieval and management of the documentation, are not sufficient to face the increasing volume and complexity of the technical documentation.

- electronic documentation systems can offer very good solutions to documentary problems :

  . easy and fast access to a complete and up-to-date centralized documentation,
  . efficient self-documentation in having both the references in the data base and the original documents in the image base,
  . cost-effective storage and management of the documentation, including security problems.

The SARDE project is only a medium-term step and want to prove :

- technical and economic feasability of such a system, and specially in our Telecommunication context,
- users' acceptation of an electronic documentation service, and the avantages they can get from it.

2. TECHNICAL DETAILS

The global architecture of the prototype is given in figure 1. The system is composed of five subsystems : acquisition, image storage, retrieval data base, network and remote display workstations.

2.1 Acquisition subsystem

The acquisition subsystem is able to take any document, in paper, microfilm or already electronic form, and to process it to be suitable for storage on optical digital disks for later display on workstations.

At this stage, the priority is given to microform and aperture card support because it is the most common in our documentation. High resolution scanners are used to digitalize an AO format page at 200 pixels-/inch (and any smaller format at 200 to 400 pixels/inch).

The restoration of image quality is of great importance for us : an efficient binarization algorithm is able to deal with high variation of contrast and damaged images.

Normalized CCITT group 3 and group 4 compression schemes can be used : bidimensionnal group 4 gives typically a compression ratio of 12 on A4 technical notes and up to 30 on larger drawings. A special method, developped in the SARDE project, involving a combination of symbol matching, lines recognition and bidimensionnal runlength coding, offers a compression ratio 5 times better. That is to say that we can store 5 more images on an optical disk, and send it in 5 less time through the network, than with CCITT group 4 methods.

An integrated coding scheme is then applied to compressed fascimile data, word-processing and CAD data (if there are some of them). The recommandations of the CCITT on document architecture will be applied ; any mixed document can be introduced in the image base.

2.2 Image storage subsystem

The image storage subsystem is build with a set of THOMSON GIGADISC optical data disks, driven by a storage machine.

The interface with other subsystems is an SM90 (French micro-computer), which :

- receives images from the acquisition subsystem (2 Mbits HDLC links),
- receives demands from retrieval data base subsystem,
- sends images to the network subsystem (64 Kbits and 2 Mbits links).

The elementary storage machines are also SM90, linked to the precedent through 2 Mbits links. Each of them has :

- enough memory to store one or two compressed images,
- a local disk to store access paths to images on optical data disks,
- several SCSI interfaces, each of them supporting one ore more optical disks controllers, with one or more optical disk drives.

Such a modular configuration is able to support a large number of optical disk drives or juke-boxes, with controlled performances-tuned to the network throughput rate.

The software service provided by the storage machine is simple but efficient and reliable :

- any image stored in the subsystem is given a numeric identifier, which is a simple mean to access this image from an external retrieval data base,

- the subsystem is transparent to data type, and other data than image data can be stored in the same way on the same media,

- reliability of the storage process is completely insured by the sybsystem, with internal checkpoints and recovery mechanisms (with help, of internal magnetic discks).

2.3 Retrieval data base subsystem

Althrough this subsystem is more classical, it is important because the users always use it to search documents, and the ergonomic quality of the whole system depends strongly on it.

The retrieval functions of the data base are a combination of three classical means of searching documents :

- fuzzy search, like in bibliographic data bases,
- navigation in a network of links beetween documents, from general to particular, which is very well suited to our technical documentation,
- historical search giving the most recently modified documents.

The workstations have the capacity to store, process and format a certain amount of data, so the queries correspond rather to an extraction of a sub-database than to a classical interrogation.

High performance, that is to say short response time and high amount of processed queries, is mandatory. To achieve this goal with a modular and not too exepensive system, a database machine approach is used (see reference [2]).

2.4 Network subsystem

Our application requires a nation-wide network, to serve several thousands potential users. The near future development of a French 64 Kbits network (RTC64), first step of the ISDN, is the main reason of our choice of 64 K bits links at the user's level.

The network access to the image database can be from 64 Kbits to 2 Mbits par second, according to multiplexing capacity of the network. A front-end machine interfaces the image database and network.

The protocols involved in image transmission are X21 at physical level, X25 at network level and Architel (national recommandations following the international normalization) for upper layers.

The retrieval databases are accessed through the TRANSPAC packet-switching network, which is widely used in France.

Adaptations to local area networks are possible if necessary, but are not a part of the SARDE project.

2.5 Workstation subsystem

The consultation workstation is one of the most important part of the system, because it is the interface beetween the end user and the system.

The workstation is a general-purpose SM90 microcomputer, with specific equipments which permit to receive the documentation through the network, to read it and to print it if wanted. Specific software services intend to give to the users easy and fast access to the distant documentation. Although these specific services provide a consultation oriented tool, it is possible for the workstation to support many other services.

The specific equipements are (see figure 2) :

- a very high resolution screen, of A3 format : this choice is done because a lot of ou documents are A3 or A2 drawings, horizontally oriented ; very high resolution means 4 million displayed points,

- a local disk, to store temporarily compressed images and database extractions,
- a dedicated processor to decode and display the compressed images,
- a simple and cheap raster printer,
- a 64 Kbits network interface,
- keyboard, mouse or joystick.

3. CONCLUSION

This paper presents a nation-wide electronic document delivery system including optical digital data disks for its image storage subsystem. A documentary workstation, part of the SARDE project is shown at the Table Top Exhibit associated to the conference. SARDE is a pluri-annual project which schedule is as follow :

- 1980 to 1982 : preliminary studies
- july 1982 : official beginning of the project.
- 1982 to december 1984 : realization of a complete prototype of the central system and of several remote workstations,
- year 1985 : experimentation in real Telecommunication context,
- spring 1986 : end of SARDE project, preparation of decisions and actions for the installation of an operationnal system.

4. REFERENCE

[1] B. GONDRAN : "PAPYRUS : un serveur d'images évolué dans le contexte d'un service de documentation" in Burotica 84, PARIS, 1984.

[2] P. PENNY, M. PICARD : "Application of novel techn ologies to the management of a very large data base" in 9th Int. Conf. on VLDB, Florence, 1983.

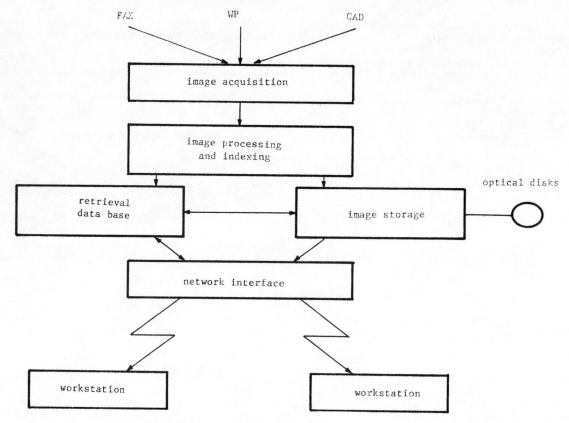

FIGURE 1 : THE SARDE SYSTEM

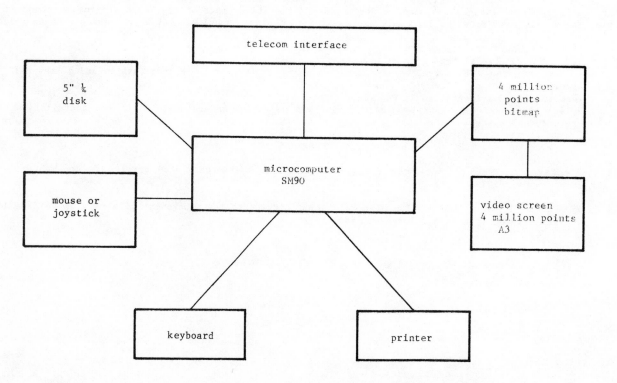

FIGURE 2 : THE SARDE WORKSTATION

Usage of an electronic document storage system for a press clipping
documentation bank

Hans-Joachim Lienau
Karsten Hartmann

Gruner + Jahr AG & Co., Postfach 2o 3o 4o, 2ooo Hamburg 36

Abstract

One of the world's largest and technically most advanced press data banks is being run as a back-up system
for journalists by Gruner + Jahr publishers in Hamburg. The text bank alone now contains 2 million pages of
facsimile text, each of which can be electronically located and retrieved within a few seconds from automated microfiche reader-printers. Daily input to the system is about 1,ooo pages, output 2,8oo. Within the next
3 years the total archive since 1972 is to be converted to electronic storage. The archiv then will contain
about 4 million pages. Texts cover all aspects of life, and over 2oo top international sources. Digital optical recording (DOR) upgrades the performance of this system. In particular: 1. extend storage space, 2. speed
up soft document delivery, 3. fully mechanize document print, 4. improve readability, 5. open up almost unlimited possibility of arranging new document collections for distributed use, 6. extend the range of document
delivery to distant user places.

Installation of the first phase started at the end of march.

Introduction

For most people the word archive suggests something dusty and antiquated, a view which seems to be inex-
tricably bound up with the image of endless shelves, documents and rows of files. Add the slightly greying
archivist and you have a scene painted by Spitzweg of engaging tranquility, seclusion and activity which is
difficult to comprehend. It is equally certain, however, that almost everyone has at some time made use of
the work which is performed in archives and has at such moments appreciated and admired it. Pupils, students,
academics, journalists, a large number of people in professional situations and interested laymen utilise
the service offered by archives and find themselves having to rely on this organised form of memorisation.

Archives perform the indispensable function of acting as a collective memory. Knowledge converning the history and the culture of a nation is kept alive by them and in this way continuity and creativity are also made
possible.

In the modern industrial and information society, however, another consideration has now assumed much
greater significance. If, as former Federal Chancellor Helmut Schmidt said, the future of the Federal Repu-
blic really does lie to a major extent in the export of blueprints, the economic aspect of information and
documentation should immediately become apparent. As a productive factor science, 'know-how' in the broad
sense, presupposes a properly functioning system of information and documentation (I&D) so that at all times
there is sufficient guarantee that the existing level of (international) knowledge can be presented. This
is the only way of avoiding unproductive multiple developments, re-discoveries of the wheel.

The ever-increasing abundance of information has frequently been described and deplored. Those who are interested are inundated by a torrent of - what is still for the most part printed- information, the huge scale
of which makes it almost impossible for any individual to sift out what is important for him. The necessary
response to this situation could and can only lie in the professionalisation of the collecting and provision of information, with use being made of the most up-to-date processing technology.

These general remarks naturally apply to all types of archives, even though it is possible and necessary
to distinguish between them according to their social and economic importance. The problems referred to in
general terms here will be described in detail below in relation to the Gruner + Jahr text documentation centre and a possible solution will be discussed.

New Methods of facsimile storage based on the
use of optical discs in the press documentation
department of Gruner + Jahr AG & Co.

1. Initial situation

From about 197o the printing and publishing firm Gruner + Jahr has gone to great lengths to create an
efficient press documentation centre, i.e. one which satisfies the requirements of its users, for both
texts and pictures. A few of the structures of this documentation centre which are of fundamental importance will be described here in order to clarify the process leading to a decision in favour of the use
of an electronic-optical facsimile storage device (optical disc). As its use will be restricted in the
foreseeable future to text documentation, the following remarks will refer to this area.

1.1. Nature, scale and organisation

Nature, scale and organisation of the stock of the text documentation centre.

As a relatively young publishing firm Gruner + Jahr does not have press archive which has been built up without interruption over the years. Rather, it is the case that various stocks which differ from one another very much in terms of content and organisation were comiled from seperate archives in the 196os and were kept and extended in a makeshift fashion.

Stocks originating from the archives of 'Stern', 'Die Zeit' and 'Constance' still clearly consti- tute the main parts. On the one hand this represented a considerable weakness, which could be seen directly in the quality of the service to be provided; on the other hand, however, this situation also offered an opportunity to seek and also to pursue new ways of performing the task required of the archive relatively free from encumbrances from the past.

There are three basic elements characterising the route which was adopted at that time:

a. The provision of a new, integrated system for the text documentation centre - laid down in the 'Index for text analysis and text research' - designed so as to be suitable for the use of an electronic search system.

b. The adoption of an Information Retrieval System tailored to the needs of a press documentation centre, i. e., the operation of a databank for administering and for retrieving documents.

c. The storage of documents in facsimile quality in a suitable mass memory system; this task was performed by the microfiche, to which in suitable equipment there is semiautomated direct access.

The databank now contains approximately 1,05o,ooo documents with approximately 1,6oo,ooo pages. In all, the Gruner + Jahr text documentation centre has more than 7,5oo,ooo pages of press cuttings, of which approximately 5,000,000 pages were allocated under the system which is operative at pre- sent. Since 1972 all the material has been made up in a uniform manner in the DIN A4 format and has also continued to be collected in files in paper form parallel with micorfilming.

1.2. Weaknesses of microfilm

From the point of view of the user of a press documentation centre microfilm has considerable weak- nesses which make it unsuitable as permanent solution to the typical problems encountered there. The basic arguments on which this assessment is based will be briefly discussed below:

a. The quality of the filming and hence of the reproduction directly depends upon the condition of the original document. All those who work in press documentation centres are aware of the fact, however, that daily newspapers in particular cannot meet very exacting standards in this respect. If anything, the standards have even been lowered in recent years. As there are limits to how much a film can be manipulated in registering half-tones, the situation today is that a relatively high proportion of indistinct and scarcely legible pictures and reproductions must be expected.

b. As well as quality, the way in which work is organised in a press documentation centre makes speed a main requirement. Microfilm of newspaper cuttings therefore has to be kept directly accessible. All the necessary equipment available hitherto operates for technical reasons with semidry copiers, which, on the one hand, require relatively long reproduction times (approximately 6 seconds) and, on the other hand, if anything, make the shortcomings in the quality even worse.

c. The need for direct acess in suitable mass storage systems ultimately means that the familiar problems of space cannot be permanently solved either. The installation of the mass storage devices requires areas offering free access and hence considerable space, which almost comple- tely offsets the gain made from what is in fact a very compact means of recording.

While, therefore, microfilm could be used temporarily in the Gruner + Jahr text documentation centre, it could not be regarded as a permanent solution, the more so since to be on the safe side it was not possible to do away with the continous collecting of the paper.

All in all, it can thus be said that in the 1970s microfilm undoubtedly represented the only possible method of running a databank in hybrid form and was capable of meeting the basic require- ments.

Quality, speed and not least costs now make it imperative, however, to look for a feasible alter- native.

2. Criteria on which the choice was based

An alternative was sought by the documentation centre relatively soon after the introduction of micro-film. From about 1978 a systematic study of the market and the available technology was carried out in order to find a method which:
a. does not have the weaknesses of microfilm described above
b. fits in smoothly with the organisational principle of the text documentation centre
c. is more favourable in cost terms than microfilm in the medium-term and the long-term
d. is reliable in terms of the future in so far as it can be adapted relatively inexpensively to new developments which cannot yet be foreseen.

2.1. Facsimile storage

Fundamental considerations resulted in the decision to continue to store and issue a facsimile in the text documentation centre and to retain unchanged the principle of a hybrid databank.

At present there is absolutely no way of knowing when, if at all, any specifiable part of the sources which are to be processed will be available in electronic form, for instance as photosetting tape. Quite apart from this there remain the questions of rapid availability and authenticity, which it is certainly not yet possible to answer in the affirmative.

Mention should also be made in this connection of the fact - to which we attach great importance - that the users of the text documentation centre are accustomed to working with documents in paper form, from which the presentation and positioning of the specific document in the printed source can be imme-diately seen. Even if the inertia of the journalists can be overcome, a significant loss in information is to be feared here as a result of the completely different method of representation.

However, the key objection from the point of view of documentation remains the fact that there are no apparent methods of automatic indexation which are practicable and economically feasible. It would not be sensible to go in this direction unless they existed.

Given these conditions, which continue to make it imperative to index each document, the method to be adopted should therefore be capable of storing and re-issuing a homogram of any given document. It can be seen from the diagram showing the way in which the text documentation centre is organised (fig.1) where the new method will be used. It is intended to replace microfilm and, in the medium-term, the whole paper archive (old stocks).
This has various advantages:
a. The user (i.e. the text documentation centre) does not have to change the whole way in which work is organised; the effects are not too complex to be understood and there will presumably be greater acceptance of the new method among employees than with other possible solutions.
b. The supplier can adapt his system to a clearly defined interface connecting all known and proven documentation systems with the new storage system.
c. There is ultimately the possibility of a modular design if the new method is required to deal with quantities of a given level.

Overall it will thus be possible to avoid over-burdening the system with too many different demands and the feasibility of using a new technology will be considerably increased.

2.2. Cost development

Given a constant level of activity in the text documentation centre, it can be assumed that - in terms of a fairly long period of about ten years in each case - there will be an above-average rise in costs, i.e. they will increase at more than the rate of inflation. This can be attributed mainly to the tremendous increase in the need for space. Compared with this, the additional costs of a higher staffing level as a result of the continually growing total stock which is more difficult to handle are almost negligible. Internal estimates showed that, under these conditions, the costs in the period 1982 (index = 100) would grow to an index value of 170 in 1992.

From this it followed that the system to be adopted should provide very compact storage, while the necessary space should clearly not develop in proportion to the expansion in quantity. Careful checking of the cost outline resulted in the initial stage in a definite preference for an electronic memory system.

2.3. Future safeguards

Two criteria were uppermost in our minds. On the one hand, a technological solution was sought which would be compatible for an unlimited period with the mass memory systems now being developed. On the other hand, the stored data was to be capable of being transmitted an displayed again without difficulty via already existing or future networks.

These requirements can only be met by a digital electronic memory system, for which a compact and not too expensive information carrier has to be found which also meets the handling requirements.

2.4. Quality safegards

The problems of quality in relation to display and reproduction already discussed with regard to microfilm could likewise, to the best of our knowledge, best be avoided with digital electronic mass memory technology. This is particularly true of the practical application with which we are concerned here, which involves only the storage of B/ W documents.

Because each point which is to be represented has to be defined as either black or white it is possible to minimise half tones, which with film are particularly troublesome in this case. One thus obtains a picture with very good contrast, which facilitates the legibility of texts.

2.5. Conclusion

The reasons specified above meant that the only possible choice for the text documentation centre was an electronic storage medium for recording digitalised information. The result of thorough studies of the available suppliers was that for reasons of handling and costs an optical storage medium ("optical disc") was the best solution. As far as the text documentation centre is aware, this medium is at present supplied as a total system ready for use only by the firm Philips. The system is known as Megadoc and the actual memory device is the DOR optical disc (DOR = Digital Optical Recording).

In April 1983 a contract was concluded between Philips and Gruner + Jahr on the installation of a Megadoc system known as the EDS (Electronic Document Storage) project.

3. Planned introduction of the optical disc system (Megadoc/EDS)

From April 1984 a pilot installation of the optical disc system will be introduced in stages by Philips in the Gruner + Jahr text documentation centre. The installation will be expanded in four phases.

a. Pilot installation in the narrow sense - in this period, covering a period of three months, the technical interaction of the individual components will basically be tested and put into steady operation. Simulated material will be used for the most part during this phase; actual production will not yet start.

b. Retrieval installation - in this phase a complete input and output configuration will be constructed with several workstations. It will be used to test, over a maximum period of nine months (July 1984 to March 1985 approximately), the efficiency of the system under conditions of production, parallel to the microfilm system which will be continued during this period, on the basis of criteria which have already been laid down.

c. Conversion installation - additional input equipment will be made available at about the same time as the retrieval installation so as to introduce substantial quantities into the system form the old stocks. This makes it possible both to test intensively the acquisition side of the system and to assess on a broad basis what quantities the overall system can cope with.

d. The enhancement of the retrieval system to the full requirements of Gruner + Jahr. As the conversion progresses, more retrieval workstations and storage capacity (jukeboxes) are planned to be added to the systems. During this stage the microfilm system will gradually being stopped.

This procedure was developed jointly with Philips over the last three years and laid down in the form of a detailed phased plan within a large-scale project study.

This study now serves as the main working document for the internal preparatory work in the Gruner + Jahr text documentation centre and for the hardware and software development which is taking place or has already been concluded in Philips.

Principles of operations

4. The input procedure

Several steps are to be distinguished:

Step 1. Preparation

Daily the Gruner + Jahr recherchers select some 800 articles from all kinds of sources: this material is cutted and pasted to A4 sheets and then stored in binders, on each article an identification number is written just before it is stored. Every binder belongs to one of the about forty special interest areas, while the complete set of pressclippings is called the current topic file.

After 6 till 8 weeks, or after a major topic (e.g. the Falkland crisis) has been closed, the rechercher, responsible for a particular interest area, selects this material in the binder again and then he decides which has to be stored in the archive, while the remainder is put in the wast paper basket. For each article the rechercher prepares a short description, following for this purpose. One particular item of this description is the above mentioned identifier.

So, in this way a batch of pressclippings is prepared for achrivation and by the nature of this procedure all articles within that batch belong to one specific interest area.

Step 2. Scanning

All documents of the batch are fed through the image scanner (a Megadoc device which makes a kind of digital electronic image of the original A4 page: sometimes the word "homogram" is used for this result) and the homograms are stored on a magnetic disc device in Megadoc.

Step 3. Identification

Next all homograms are recalled from the magnetic disc and displayed on a screen specially designed for this purpose, the so called High Resolution Display. The electronic copy can now be compared with its original to check the qualitiy. In some cases a re-scan may be required. At the same telling the Megadoc system that this particular homogram is further referred to by this identifier.

Step 4. Archivation

When these actions have been completed for the whole batch, all homograms are written by the Digital Optical Recorder on the glass disc. Batches from different interest areas are put on the disc in their scanning sequence. Later on, every batch is written again, but now to a final place in one of the 60 till 100 discs, which are organised by interest area. As the initial disc is to be used for back-up purposes, the original pressclippings can now be thrown away!

Step 5. Indexation

Although all homograms are available in the Megadoc jukebox cabinet, there is at this point in time only one way to retrieve one, namely by entering the originally issued identifier, provided one still remembers this number.

An accepted way to overcome this problem is to apply a piece of software called "Information Storage and Retrieval. Gruner + Jahr has developped such a tool already some years ago taking into account the specific requirements for press documentation environments. This programme is implemented on a large timesharing mainframe and recently it has been adapted to interface Megadoc by a standard datacommunication protocol.

The short description prepared by the rechercher (see step 1 "preparation") including the identifier is keyed in on a terminal of this mainframe and the index with inverted files of the retrieval system is updated accordingly. Now the identifier provides the link between the Information Storage and Retrieval system and Megadoc.

Principles of operations

5. The retrieval procedure

If an editor or a journalist of one of the Gruner + Jahr periodicals requires information, this request is received by a recherche member of the documentation department. The rechercher has an abundance of information at his disposal, among else the homogram collection in Megadoc. The request of the editor ("Give me all information on the comments of Mr. Reagan with regard to the Falkland crisis") is interpreted by the rechercher and keyed in on a terminal of the Gruner + Jahr retrieval system on the main-frame. Applying the indexes etc. the system finds and displays a number of references to pressclippings.If the rechercher is satisfied the identifiers are sent over the data communication interface to Megadoc which reads the appropriate homograms from the optical discs. The homograms are either displayed on the special screen or printed such that hard copies can be forwarded to the editor.

6. Prospects

The Gruner + Jahr text documentation centre considers that the introduction of this new mass memory will have another fundamental advantage in addition to those already discussed. The digitalisation of information in principle creates unlimited possibilities for transmission. In other words, by making use of facilities for the teletransmission of data it will be possible to develop a documentation system which is available in all places from what was hitherto a largely fixed overall system.

This already finds expression in the first phase of development in that documents will be issued directly via any telephone in a telephone circuit.

Not only, therefore, is a basic problem of all databank suppliers - rapid provision of literature - much nearer to being solved, this also has very practical significance for the day-to-day work of the text documentation centre, which will be able for the first time to provide the numerous external editorial offices of the publishing house promptly with the required documentation.

As it is guaranteed that this system is capable of expansion, also in terms of coping with considerably larger quantities, we anticipate being able to achieve a long-term, future-oriented solution to the main problems of a press documentation centre.

ORGANISATIONSSCHEMA
DER
G+J-TEXTDOKUMENTATION

Fig. 1

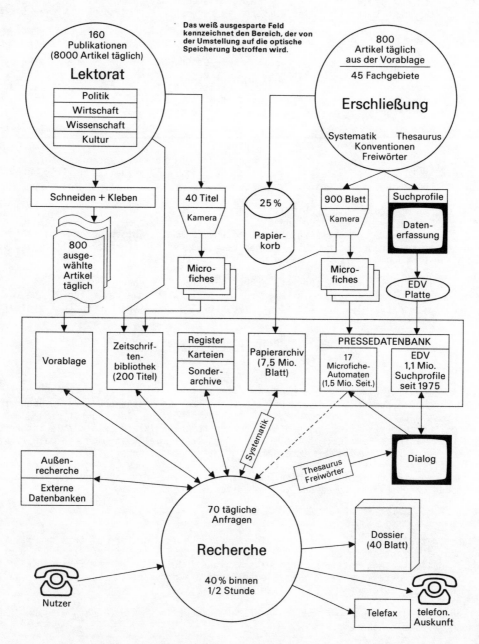

A layered storage structure for images confronted with the use of X-ray images in a hospital

R.C. van Rijnsoever, J.P.J. de Valk, A.R. Bakker

BAZIS, Leiden University Hospital
Rijnsburgerweg 10, 2333 AA Leiden, Netherlands

Abstract

In modern hospitals the attention paid to the problem of lending and archiving of images is increasing. In almost every (big) hospital complaints are heard concerning missing X-ray images and lack of storage space. The use of the computer and digital storage systems might lead to a solution of these problems. However enormous amounts of data are involved, ranging per image from 128 kbytes for a CT-scan to over 2 Mbytes for a thorax X-ray. For the Leiden university hospital (940 beds) this implies a daily production of about 8000 Mbytes of uncompressed data after digitisation.

At this moment, a study is in progress carried out by the central development organisation (BAZIS) of a group of Dutch cooperating hospitals, aiming at a well founded choice of a system framework. During this study extensive use is made of the information available from the existing Hospital Information System (HIS). The acquired data will be used as input to a simulation model, which should give more insight into the behaviour and necessary characteristics of the memory configuration required. This configuration will consist of different layers.

Introduction

In a hospital much information is obtained with the use of diagnostic images. The radiological department is by far the biggest producer of images. At present X-ray images are mostly photographs made by convential methods, although the use of techniques yielding digital images is increasing (computed tomography, ultrasound, nuclear magnetic resonance, nuclear medicine etc). In a large hospital (1000 beds) with many outpatient visits, yearly about 130,000 analogue X-ray examinations are done, requiring about 30,000 square meters of film. In addition about 16,000 examinations resulting in digital images are done, generating about 50 Gbyte of uncompressed data[1]. Normally, a hardcopy is made from the available digital data for use later on. This hardcopy may display less than 5% of the original digital data[2].

In the Leiden university hospital the results of every examination are stored in a jacket which, in its turn, is stored in an archive. The total contents of the archive comprises over 2500 running meters of jackets stored in 7 different buildings. Various reasons may underly the request of the X-ray history of a patient, e.g. the visit of a patient to a specialist or a new examination. The retrieval of pictures belonging to old examinations may be a time consuming task. Sometimes the pictures are 'somewhere' in the hospital or in a part of the archive in another building. In some cases photographs are even missing.

Apart from these problems there is the uniqueness of the pictures; a picture can only be in one place at a time. If, for instance, a patient visits a specialist, the specialist may have asked for the X-ray history. If the specialist thinks a new examination to be necessary, he sends the patient to the radiology department where the radiologist will want to look at the patient's X-ray history. After the examination the patient returns to the specialist who probably wants to see the history again to compare it with the results of the examination just made. It would be much more convenient if both the specialist and the radiologist would have a copy of the patient's history.

In the Leiden university hospital the IMAGIS project[3] has been started to study the various aspects of a Picture Archiving and Communications System (PACS), coupled to the existing Hospital Information System (HIS). The part of the study dedicated to the problems related to the archival and retrieval of diagnostic images will be described. With the progression in storage technology it might become possible in the future to store the results of all examinations in a digital storage system. Several parameters will affect the performance of the storage system. It may be evident that determining the optimum values of the parameters is a difficult task, partly because few parameters, if any, are independent. The method used in the Leiden university hospital in an attempt to solve this problem, is to describe the storage and retrieval system in a model. For every component of the system a process description is made, which is used in a simulation program. Knowledge about the influence of the various system parameters, which is necessary for the optimum realisation of a digital storage system, can be obtained using the simulation.

A digital archive

The implementation of a digital storage system where the results of all examinations are stored, would bring the following advantages:
- Fast access to the results of every examination.
- Easy access to particular classes of images. Apart from the access to all images belonging to one patient also (e.g. for training purposes) to images with a specific property.
- Digitally obtained images can be recorded (and reviewed) retaining their full information content.
- Images can be reviewed at several places simultaneously.
- Original images stay in the central archive and therefore can not be lost.
- Reductions of cost may be expected, due to film (=silver) and space savings.

Apart from archiving advantages we would like to mention the following possibilities:

- Image processing techniques can be applied in order to improve the diagnostic image quality (e.g. noise reduction, contrast enhancement etc).
- The quantification of results can be improved (e.g. concerning a specific disease). By the use of standard techniques an objective 'measurement' can be realised which can be useful for comparison purposes in particular.

Some major difficulties to be faced are:

- A change of working situation will have to take place.
- Feasible systems will be quite complex (e.g. a very high speed broadband network is needed).
- Very high resolution display stations will be necessary due to the high information density of the currently analogue images which are to be digitised.

Required storage space

For storage of all the examination results in a digital archive a very large capacity is needed. Although for an analogue X-ray photograph a spatial resolution of 400 pixels per square millimeter is considered to be an upper limit[4] beyond which no improvements in detail are to be expected, diagnostic information will surely be contained in a resolution of about 64 pixels per square millimeter and 256 grey levels. This implies an annual generation of over 2000 Gbyte of image data, following from the 30,000 square meters of film, produced in a large teaching hospital, as mentioned before. Although at present every examination result remains stored regardless of age or use, an archiving period of 10 years is considered to be sufficient , at least for the data which must be retrievable with reasonable speed. This implies a total needed storage capacity of over 20,000 Gbyte.

Of course compression techniques exist[5] which may reduce the needed capacity but the techniques involved should not be overrated. Many so called data compression techniques in fact are data reduction techniques. The difference is that with compression no original data is lost, thus the process is reversible, whereas with reduction data is actually lost which makes it impossible to restore the original picture completely. For many purposes reduction is very well applicable (e.g. facsimile) but with X-rays one should be very careful not to loose essential information. Since only a specialist is able to determine whether important information is lost, it is very important to involve the radiologist when working on reduction techniques. A reduction technique which may be useful is where the radiologist indicates which regions can be stored with lower resolution and which regions are not interesting at all for diagnostic purposes and thus can be deleted (e.g. an X-ray image of a leg; the areas around the leg will usually not contain important diagnostic information).

At present a compression factor of about three can be expected, yielding in our case the need for an archive capable of storing about 6,000 Gbyte. Today no storage device is able to store and automatically access such a huge amount of data on-line. However if human intervention is allowed a solution with for instance optical discs might be considered. This would require an archive of at least 3,000 discs, based upon a capacity of 2 Gbyte per disc. Most of the discs would be stored off-line in racks, from which an operator could take them for mounting on a digital optical recorder. It should be realised that storage technology is still improving. Storage media with higher capacity can be expected to become available.

A possible configuration with multiple layers

In order to reduce the waiting time after a request to acceptable values, the memory configuration should consist of different layers. At least three layers will be necessary.

- The first layer would be the archive, in which the results of every examination are stored. Since this implies a layer containing over 6,000 Gbyte of image data, the accesstime will be long. If the archive consists of optical discs the accesstime will be the time needed to get the disc out of the racks containing the 3,000 discs, plus the mounting time of the disc on the digital optical recorder (DOR). When an X-ray history, consisting of the results from several examinations, is requested, the total accesstime will depend upon the storage method applied. If the examination results are stored per patient the accesstime of the history is the same as the accesstime of an examination. Storing per patient implies that the storage should be organised in a way that in general more images of one patient can be stored together. This requires a significant additional amount of capacity.
On the other hand, if the examination results are stored in chronological order, the total accesstime increases with the number of examinations in a history.
Of course, if the examinations are stored using a compression technique, the decompression will probably influence the time needed for the retrieval of an image.

- A second layer is necessary, which acts as a buffer. Ideally, examination results with a high probability to be requested, are stored in this layer. This layer can be filled partly during low activity hours (e.g. at night). In this way activity is spread over 24 hours instead of over 8 hours. Examination results can be automatically transferred from the first layer to the second, if an adequate link to the HIS exists. In the HIS the appointments of the patients with a medical specialist will be registered. Prior to the visit of a patient, (part of) the history of the patient can be transferred from the archive to the second layer, under control of some background program. When the radiologist actually requests the history, the history is available on a higher level, which will result in a reduced waiting time.

- The third layer is the local storage in the working station, in which the working set (e.g. the history of a patient) of the user is stored and for which very short waiting times are demanded. Minimum waiting times are necessary because of the way the data is used. Images will be compared with others. During the interpretation process images will be replaced by others, which must occur instantly so as not to slow down the interpretation process. If the response time coupled to a request for a picture from the working set is more than a few seconds, the system will no longer be regarded as userfriendly by the clinician and acceptance of the system will become more difficult. After all, the change from the current use of a lightbox to the digital working station will be hard enough[6].

Performance considerations

The performance of the system described above is dependent upon several parameters, such as:

- The capacity of the digital optical recorder(s) and the buffersizes of the second and the third layer.
These parameters directly influence the performance of the system. If the second layer has a very big capacity, there will be a minimum of requests to the archive (first layer), which improves performance. Similar considerations apply to the capacity of the third layer.
- The accesstimes of the various layers.
In particular for the third layer the accesstimes should be short.
- The transferrates of the diverse layers.
Due to the large amounts of data to be transferred the transferrates should be high. With an X-ray photograph of 35x35 cm about 8 Mbytes of digital data will be involved at maximum. If waiting times of less than one second are demanded in the interpretation process, when only the working set is used, the transferrate of the third layer should then far exceed 8 Mbytes per second. For the first and second layer the transferrate is more difficult to determine due to the influence of the other parameters on the waiting times.
- The configuration of the network installed and the network protocol adopted (preferably standardised).
The network should be able to cope with the transferrates of the different layers and the accesstime to the network should of course be sufficiently low. Furthermore it should provide means for error detection and recovery. Many configurations are possible like a ring or a star or a fully interconnected network.

- The search method applied and the datapath followed.
  Depending upon the accesstimes and the network traffic the search method could have a significant influence on the search time for a picture. A request may travel from the highest level downwards until the image is found. On the other hand the bookkeeping of the whereabouts of the images in all layers could be done centrally.
  If every requested picture travels upwards from the lowest level the waiting time will be the sum of the waiting times of each layer, whereas if every picture travels directly from source to destination only one waiting time is involved. Of course, in the latter case, the layer omitted won't act as a buffer unless specific action is undertaken.
- The storage method applied (e.g. per patient or chronological).
  As mentioned before this may influence the total accesstime to a history. Another aspect is the use of compression or reduction techniques. When and in which layer decompression is executed will affect the waiting time for the images.

A simulation model

Since most parameters of the digital storage system affect each other it will be difficult to determine the optimum values. Moreover some parameters will change as technology progresses (e.g. the on-line capacity of the first layer). In order to meet future trends in technology a simulation model is constructed with adjustable parameters. In the model each component which is simulated is assigned a process description. The process description describes in detail the actions a component has to perform, related to the actions of the other components. An optimum configuration may be obtained by a systematical choice of different sets of parameter values. The input of the simulation model consists of data considering the current use of radiographs. This implies the need for knowledge of when and how many times a radiograph is requested, of the time a specialist studies the radiograph etc. Part of this knowledge is obtained by using the information currently available in the HIS. In this way we know that on average about 85% of the patients visiting the Leiden university hospital already have an X-ray history, which consists of, on average, the results of 13 examinations. The mean number of examinations performed per visit is 1.2. Almost 500 examinations are done each (working) day. On average the results of an examination occupy 0.3 square meters of film (19 Mbyte of data when digitised) which implies a film area of 4 square meters (255 Mbyte of data when digitised) per history. So each day 125 square meters of film (8 Gbyte of data when digitised) are generated.
It has been determined which part of the examinations is for inpatients. For outpatients the appointment period is being measured, leading to input for an algorithm to activate images at the second level.

Conclusion

The data mentioned above are part of the data which will be used to simulate a model of a layered digital storage system for diagnostic images in a hospital. The results obtained may be used for a limited scale PACS experiment with images already in digital format (CT ultra sound, NMR). Because of the lower spatial resolution of these images, the amount of data is small, compared to the data contained in a (chest) radiograph after digitisation. Currently available storage technology seems able to cope with the total amount of already digitised data involved.

References

1. Kouwenberg, J.M.L., Cuignez,F.J.G., Automating Image Data to On-line Patient Information, Proceedings of the Fourth World Congress on Medical Informatics Amsterdam MEDINFO'83; Eds. J.H. van Bemmel, M. Ball and O. Wigertz, IFIP-IMIA; North-Holland Publ. Co., (1983), 373-376.
2. Maguire, Jr., G.Q., Zeleznik, M.P., Horii, S.C., Schimpf, J.H., Noz, M.E., Image Processing Requirements in Hospitals and an Integrated Systems Approach, Proceedings of the First Conference and Workshop on Picture Archiving and Communication Systems (PACS) for Medical Applications, Newport Beach, Cal. Editor A.J. Duerinckx, SPIE Vol. 318 (1982) 206-213.
3. Bakker, A.R., Kouwenberg, J.M.L., Retrieval and Representation of X-ray Images, a Layered Structure, Proceedings of the ADV, 7th International Congress on "Datenverarbeitung im Europaischen Raum". Wenen, (1984), in press.
4. Arink, G.J., Electronic Image Management for Medicine, Proceedings of the First Conference and Workshop on Picture Archiving and Communication Systems (PACS) for Medical Applications, Newport Beach, Cal. Editor A.J. Duerinckx, SPIE Vol. 318 (1982) 410-417.
5. Netravali, A.N., Limb, J.O., Picture Coding: A Review, Proceedings of the IEEE, Vol. 68, No 3. (1980) 366-406
6. Seeley, G.W., Stempski, M., Roehrig, H., Nudelman, S., Capp, M.P., Psychophysical Comparison of a Video Display System to Film by Using Bone Fracture Images, Proceedings of the First IEEE Computer Symposium on Medical Imaging and Image Interpretation ISMIII'82; CH 1804-4, W-Berlin, (1982), 212-216.

Electronic Information for Primary Care

Maurice H. King

Kenyan/German Project for Hospital Technology, P. O. Box 19, Nyeri, Kenya

The problem

Few sciences have more information problems than medicine. The student has an ever increasing amount to know. Some of them have even complained that they have "so much to remember that they have no time to learn." Once qualified, a doctor finds it increasingly difficult to keep up to date, particularly in clinical pharmacology.

Summary of proposal

This is the information age. So it is proposed to bring the new technology to the assistance of the general practitioner (to begin with), and to record electronically, probably on a videodisc (or discs) anything he might need to look up. The aim is to make it as easy as possible for him to do this in the presence of his patients, and if necessary invisible to them - both for their immediate benefit, and for his own continuing education.

A videodisc is about the size of a large gramophone record and has the information storage capacity of at least 64 large books. One of these discs will be distributed to subscribers once or twice a year by mail order as a form of "electronic journal", or "electronic encyclopaedia". The information on each disc will be a combination of text, diagrams in black and white or colour, moving and still, speech and other forms of sound. The encyclopaedia will also contain clinical computer programs, which the practitioner can take off his vidoedisc and "dump" into his own microcomputer. All this information will be instantly accessible, either through a keyboard, or he might even be able to speak to his computer. The system will be interactive in that he will be able to ask and answer questions, and be lead through an algorithm. His relationship to it will thus be much more dynamic than with an encyclopaedia of the traditional kind.

In due course, and with suitable support, a highly authoritative, comprehensive "globally relevant system of clinical medicine" might be developed on one or more discs. The hope is that the encyclopaedia will have such a wide circulation that it can support a large editorial team and yet be cheap. The aesthetic potential of the med'ium is so great that it should be possible to produce something really beautiful. Conceivably, the encyclopaedia might become something that no doctor anywhere could feel he could possible be without, and radically alter his interaction with information both before and after qualification.

It will above all be practical, and systematic, "user friendly", problem oriented, organised, and cross-linked. It will serve as: (1) a source of information during a consultation, and (2) as the means of further study and continuing education afterwards. The information in it will appear on the same screen that many practitioners already use for their records. The project will however not be primarily concerened with the records aspect of computer use, except to ensure hardware and software compatibility.

The needs of the practitioner, and the constraints under which he works in his consulting room will be studied with the greatest care. The closest collaboration will be maintained with individual practitioners, and their organisations. The project will be cooperative and "symbiotic" and will aim to assemble and systematise knowledge, rather than to create it. For example, if the clinical pharmacology described below already exists, or is under development (which it may be) the aim will be to cooperate with its developers, and encourage them to lend it to the encyclopaedia. The same principle will be applied to all other information.

The need for such an encyclopaedia is not confined to the United Kingdom. If possible, it should have a global orientation. Hence the great value of sponsorship by WHO, who have asked for this paper, and who are in an excellent position to establish colaboration rather than competition. The owners of copyright material are likely to lend it to the encyclopaedia for the common good, if WHO sponsors it. WHO is also in a good position to promote the standardisation of both hardware and software, as well as being able to promote its adaptation and diffusion in the third world.

The project is submitted by the writer, who has just started his 11th book for Oxford University Press, four of them at the request of WHO, and at least three of them best sellers. It is seen as lasting five years initially, and costing about $1.57 million dollars. In view of its novelty, it is planned to make a cautious beginning, to realise its simpler and least expensive aspects first, and to achieve its full potential gradually. It is planned in two phases, an intial pilot phase lasting three years ($0.89 million) to develop and try out the system, followed by expansion phase ($0.67 million) to expand the encyclopaedia to the point of commercial viability. When the project is successfully completed, it will be taken over by the Press as a viable self-supporting non profit making activity, to be run for the good of all, under the direction of an international board, with representation from the third world, and hopefully from WHO.

The proposals presented are based on the technological possibilities available in early 1984.

Integrating electronic and printed information

If the potential of electronic information retrieval is so great, what are its limitations? (1) Print on paper is much easier to read than text on a screen. "Ten point times Roman" and the eye of the educated reader have become so beautifully adapted to one another that print is unlikely ever to be completely superceded. By contrast, all text on a screen (so far) is crude and much less easily readable. (2) No equipment is needed to read print. (3) Print is more portable, the Lancet, for example, can equally comfortably be read in bed, or waiting in the departure lounge.

Electronic information has the advantage over print: (1) When the quantity of information to be held is very large. (2) When frequent updating is required. (3) When moving illustrations, sound or interactive computer programs are needed. (4) When there are cultural and pyschological objections to the use of a book. In most cultures, a doctor may not refer to large books in the presence of his patients. To do so is to lose status. The most he is allowed are little books like the BNF (British National Formulary) or "Mims". In some cultures (Indonesia) he is not even allowed these, and is expected to know all doses by heart. By contrast, there appear to be no such prejudices to the use of "a computer", which may actually increase the status of a doctor in the eyes of his patients. As more of them become "computer literate" they may increasingly expect, and perhaps even demand it.

The challenge is to integrate electronic and printed information, so as to make the best of both of them. So it is proposed to: (1) Make it possible for any text on the videodisc to be printed out on the kind of printer that microcomputer systems are already equipped with. (2) To integrate the videodisc with the following books and journals to begin with, and to add others later.

(1) Subsequent editions of the Oxford Textbook of Medicine (the first edition has just appeared). This is published by the Press and promises to be the standard work in its field. It is a comprehensive, authoritative, inexpensive two volume work, (£45, 1.6 pence or US cents 2.5 a page of a thousand words) with a wide range of international contributors. Anything in the Textbook which a doctor might usefully want to look up in the presence of his patients will also be available electronically. For example, he might want to refer to diagnostic and management routines. To complement the Textbook, coloured illustrations (large quantities of still slides, shots of movement (characteristic gaits for example) would be in the electronic encyclopaedia together with clinical algorithms, and diagnostic computer programs etc. The general background to a particular problem, its physiology and pathology, will probably be more conveniently read in print (although it might also be available electronically). The Oxford Textbook and the electronic encyclopaedia will be cross referenced to one another, and contributors will be asked to contribute to both. The Press are enthusiastic about the possibilities of this integration.

(2) Articles aimed at improving general practice published in the "The British Medical Journal" and the "Practitioner" will be completed (many of them are fragmentary from the user's point of view) and made available electronically, in a problem oriented form. These electronic versions, will be signed by the contributors, as in the journal. The printed articles in these journals will include a note to say that they were being adapted for the encyclopaedia. The practitioner on reading the article, will then know that he need not remember the details, but will be able to call them up later electronically, if he wishes. Neither of these journals have yet been approached officially. This proposal is however being sent to them.

The textbook is academically rigorous and "up-market", The encyclopaedia will be aimed to begin with at everyday problems and will be more "down-market". ("She has just swallowed 20 Mogadons - what do I do?"). There would in fact be ample space for both orientations.

Data retrieval and self instruction

Videodiscs are already in increasing use for teaching purposes, and are generally held to be the ultimate in audiovisual methods. They are only now being adapted for the data retrieval purpose suggested here. There is no reason why a given disc should not be used for both purposes. For example, a disc will contain the basic information needed for retrieval on, say hypertension, and also a refresher course in its management. These two uses will be on separate pathways through the disc, making use of the same frames where necessary. The only constraint on this double mode of use is that the teaching mode would be mostly still frames with sound, whereas most uses of the disc for instruction involve much movement. Provided sufficient sound track can be made available on the videodisc, its capacity for self instructional material of the "slide/tape" variety will be very large.

The critical position of clincal pharmacology

The information problem is nowhere more desperate than in the prescription of drugs. Electronic assistance might be invaluable. Hopefully, the entire BNF, will be made available electronically, in an expanded form, and suitable programs provided to enable the interactions and incompatibilities of any combination of drugs to be seen. This is such an urgent priority that by itself it is probably a sufficient justification for electronic assistance.

If a practitioner could not, or did not wish to use a keyboard, he might call up the drug list using the voice recognition system described below. He could bring the cursor to each drug in turn by spelling it out quite quickly, followed by the command "comptibility", whereupon the interactions of the list of drugs would appear. Although this might seem unreal, it should not prove too difficult.

The impact of optical storage technology on image processing systems

Daniel T. Garges and Gerald T. Durbin

StorageTek Integrated Systems, a subsidiary of Storage Technology Corporation
6500 Rock Spring Drive, Suite 204, Bethesda, Maryland 20817

Abstract

The recent announcement of commercially available high density optical storage devices will have a profound impact on the information processing industry. Just as the initial introduction of random access storage created entirely new processing strategies, optical technology will allow dramatic changes in the storage, retrieval, and dissemination of engineering drawings and other pictorial or text-based documents. Storage Technology Corporation has assumed a leading role in this arena with the introduction of the 7600 Optical Storage Subsystem, and the formation of StorageTek Integrated Systems, a subsidiary chartered to incorporate this new technology into deliverable total systems. This paper explores the impact of optical storage technology from the perspective of a leading-edge manufacturer and integrator.

Evolution

Until now, processing of high-volume text and image data has been constrained by the cost-effectiveness of on-line storage. The industry has seen rotating magnetic storage evolve along two increasingly distinct paths - performance and capacity. Performance devices are characterized by relatively low capacity and fast access time, and are essential for key input-output functions inherent in today's large-scale response-oriented systems. Capacity devices, on the other hand, sacrifice access time due to the higher number of stored bytes per data path. These new devices store as much as 2.5 GBytes of data on two non-removable spindles. As these devices approach the practical limit of thin-film magnetic technology, the required tolerances will dictate continued use of non-removable media.

For the high-volume data user, system design involves a combination of performance and high capacity direct access storage, with magnetic tape as the principal backup/archiving medium. However, even with the latest high-capacity disk and high performance magnetic tape, the user must trade off on-line availability against cost of storage, and must live within the constraints of elaborate archival and restore mechanisms which result in access times measured in minutes or hours. The application must still be constrained by the limits of the technology.

Ideally, the solution to this problem would embody the random access capability of magnetic disk (at a reasonable level of performance), as well as the removeability, transportability, and archivability of magnetic tape. In addition, one would hope to surmount some of the inherent problems of magnetic tape as an archival medium, such as erasability and short shelf life, while providing access within seconds to vast amounts of data at an affordable price.

Recent announcements of digitally encoded optical storage systems, such as the Storage Technology 7600 Optical Storage Subsystem, promise a solution to this problem. The STC 7600 offers numerous advantages which meet the requirements of imaging applications:

1. Capacity - The STC 7440 Optical Media Unit (platter) contains 4 GBytes of user data on one side of a 14" disk. Through attachment to an automated library, hundreds, or even thousands of these platters could be available on-line within seconds.

2. Performance - The 7640 Optical Storage Unit provides an average access time of 85 msec, with data transfer rates up to 3 MBytes per second.

3. Archivability - The media provides a life of 10 years without loss of data integrity, with a bit error rate of 1 in 10^{13} (or one bit in error every 312 platters). In addition, the non-erasable nature of the recording media reduces backup requirements dramatically.

4. Availability - Since the media is removable, it can be made available on-line very quickly. In addition, it lends itself to transport easily, without concern for damage from temperature, x-rays, or magnetic fields.

5. Affordability - The cost of an optical solution is less than that of combined magnetic disk and tape solutions while eliminating the data volume constraints of those technologies.

Application

To understand the diversity of the image processing industry, one need only compare an electrical circuit diagram with a page from a standard insurance policy. Each represents a major segment of the burgeoning image processing market. The advent of optical storage is reshaping the cost-justification formulae for on-line storage. Documents which would never have been considered for computerization are now likely candidates for data capture. When the growth of office automation is combined with the availability of optical technology, the concept of a paperless office filing system moves much closer to reality.

The successful implementors of these new systems will have three major distinguishing characteristics:

1. Manufacturing-level expertise in both magnetic and optical storage devices.

2. A dedicated staff of systems integrators, with sufficient software expertise to ensure a working large scale system incorporating multi-vendor hardware.

3. A world-wide maintenance and support organization.

In order to give the reader a better understanding of the next generation of image systems, the next three sections discuss system capabilities and characteristics from the perspective of a leading manufacturer and integrator. Three areas will be discussed: Input, Storage Control, and Output.

Input

While many techniques exist for image input, the two most commonly used are laser scanning and Charged Coupled Device (CCD) scanning. At present, laser scanning technology has a number of deficiencies which preclude cost-effective scanning of documents or engineering drawings. CCD scanning, on the other hand, has improved greatly, and now allows high speed/high density scanning of virtually any type of input, from microfilm to 18-foot-wide hard copy. This most recent CCD technology provides the foundation upon which to build a new generation of intelligent scanning workstations.

The scanning workstation is a critical component of any image processing system. While it is unrealistic to expect any single workstation design to be universally applicable, a "standard" scanning workstation might have the following characteristics:

1. Each workstation would be a self-contained module capable of scanning, editing, and storing data. Output from the workstation might be transmitted to a central computer, or simply stored on a local device.

2. The scan rate would be operator-selectable for 100 to 800 points per inch. This is an important technological breakthrough which will allow the scan to be adjusted to the document or drawing being processed.

3. The scanner will be designed for 11"x17", two-sided capture. This will allow the front and back of forms to be processed in the same pass, and multipage forms to be handled much more efficiently than previous technology would allow. The potential exists to quadruple the throughput of existing systems through two-sided scanning of multipage forms.

4. The captured image would be immediately available for operator review, and will undergo on-line adaptive processing, compression, character recognition, indexing, or reformatting under local control.

5. Facilities will be available for hard copy, microfiche, and aperture card scanning. The data stream will be modifiable within the workstation to accomodate the various types and sizes of media.

In summary, the newly developed scanning technology will be integrated into powerful, microprocessor-based workstations. These workstations will function in both standalone and communicating mode to provide an efficient and flexible entry point for image data.

Storage and control

Once image data has been captured it is part of the information system, whether resident in an individual workstation or a central archive. The system design must provide control of the data through its life cycle of storage, transfer, and dissemination. An integrated system will make use of several different types of storage:

1. High Performance Magnetic Disk - Relatively low capacity devices with extremely fast access times. These devices are necessary to store system programs and indices necessary for high-volume multi-user environments. In many cases, a solid state device such as the STC 4305 may be used in this role.

2. High Capacity Magnetic Disk - High capacity devices such as the STC 8650, or the STC 8380, which stores 2.5 GBytes of data on two spindles of non-removable media. These devices provide the primary random-access memory for the system, and would be used to hold active data.

3. High Density Magnetic Tape - Used for conventional sequential storage and periodic back-up of magnetic disk memories. The STC 4670 tape subsystem is an example of this technology.

4. Optical Storage Units - The STC 7440 will store 4 GBytes of data on one side of a 14-inch removable platter. The platters are provided in easily-loaded cartridges for protection during handling. The recording is permanent (non-magnetic) and each platter holds the equivalent of 40 reels of 6250-bpi magnetic tape, with a bit error rate of 1 in 10^{13} over a ten-year shelf life.

In addition, optical platters could be stored in automated libraries for very high volume users. These optical data libraries would vary in size from a few hundred to over a thousand cartridges, with access to the requested cartridge in seconds without operator interaction. This approach creates an archive previously unattainable, with capacities in the 1-5 TByte range. Libraries will support multiple optical drives, so that the subsystem can be configured with greater flexibility.

The optical platter is a key element in the overall system design. Its 4 GByte capacity and compact size allow it to be used for data transfer as well as archival storage. Platters produced at standalone workstations can be transported to a central system where they can be loaded either manually or through an automated library. Users will also have an interesting new mode of high speed data transmission. Consider the following: assuming a high speed 50KB telecommunications line between workstation and central system, it would take over seven days to transmit the contents of a 4 GByte platter. Alternatively, the user could simply send the platter via overnight mail, effectively transmitting all the data in a matter of hours. The fact that the optical media is impervious to magnetic interference makes this strategy even safer than shipping today's magnetic tape.

One of the most promising features of the new design is the ability to store both intelligent (computer processable) and non-intelligent (imaging) data on the same media. This capability results from the all-digital nature of the system, and signals a new era of processing flexibility. The workstation operator will be able to process both numeric and non-numeric data, and conversions within the system may be used to achieve an optimal mix of intelligent and non-intelligent data for a given application.

The storage and control of data in future systems will encompass a variety of hardware devices, all of which must function under or be compatible with a single set of software. This raises two key issues: first, the integration of the equipment must be technically feasible; second, someone must create integrated working systems. As a leading manufacturer, STC has responded to each of these challenges.

In order to insure software compatibility among all devices, STC has standardized the optical storage unit on the IBM MVS SP 1.3 Operating System. In this configuration, the optical device will communicate through an STC 8880 Controller which attaches directly to an IBM-type channel.   Since all of the other devices discussed are also supported by IBM software, the result is complete operating system compatibility. In addition, an Intelligent Peripheral Interface (IPI) has been announced which will allow the optical storage unit to operate in a non-IBM environment.

The second challenge required an even stronger response. As the need to incorporate optical technology into complete functional systems became clear, an entire subsidiary was formed. StorageTek Integrated Systems has been chartered to integrate a variety of STC and non-STC products into functional total systems which address specific user requirements. Major components such as workstations will be standardized and used as the building blocks for customized user systems.

Output

This section discusses in general terms the complex problem of data distribution. The utility of any information system is largely a function of its ability to disseminate data in a timely and functionally useful way. In image processing systems, the problem is complicated by the high volume of data and the variety of media. With respect to volume transmission, the optical platter itself provides an innovative vehicle, as described in the previous section. For actual output, however, a variety of devices will be used to satisfy unique user requirements. Given the all-digital nature of the system, there are many existing devices well-suited for integration into image processing systems.

Hard copy output can be produced using several current techniques. Engineering drawings and documents up to "B" size may be reproduced by either laser or raster printers, and those above "B" size by raster printers. The engineering drawing is restricted in quality by the capability of the printer and its ability to image on sensitized paper. Laser printers offer high speed production capabilities and high quality images. These printers are capable of printing up to 120 pages per minute with excellent quality. Both raster and byte data may be transferred from the digital storage system directly to the laser printers for reproduction of the information onto hardcopy.

Digital data may also be routed to Computer Output Microfilm (COM) equipment for generation of high resolution images on film. In document systems this information may be generated onto standard NMA formatted microfiche or roll film. In the engineering drawing environment, the information is transferred first to indexed silver film and then placed in special film-to-card diazo duplicators that operate under control of the central computer system to output the requested number of duplicate aperture cards. It is also well within the realm of reason to consider the replacement of existing archival COM with the optical disk itself. Since the information normally written to COM is from a digital data base and the COM output process converts the digital to analog, there are no provisions to reverse this process should the data be requested again. The optical disk will allow the incorporation of the digital data into an all-digital information system without the penalties that currently exist with analog microfilm devices.

A third distribution system consists of visual displays. The display workstation is comprised of a microprocessor-controlled display with the ability to process both intelligent and non-intelligent information on the same screen. Data from local area networks, or from remote transmission is stored and displayed under complete local control. The local workstation would also have its own disk storage and printing capabilities. Since this is a vital link in the information flow process, each display station would be a workstation in itself, fully capable of interaction with the host computer or standalone operation as a remote microprocessor for additional cost effectiveness.

The combination of hard-copy, optical archiving, COM, and visual display will provide output capability to meet a variety of user needs. All of this hardware is currently available, and is being integrated into innovative system solutions.

Conclusion

The age of high technology image processing systems is here. Recent technological advances in this area now provide the tools that will allow introduction of all-digital technology into an arena that has long suffered from the constraints of analog-based design.

In addition to providing a critical technological breakthrough in the introduction of optical storage, Storage Technology Corporation has embarked on a concentrated effort to offer serious users a solution to their information processing problems. The incorporation of non-intelligent information with intelligent data from computers now allows the capture and dissemination of information that previously has been unavailable. The key to successful implementation will be the integration of newly available scanning and storage technology into customer systems which utilize standard hardware modules and are backed by a world-wide support organization. Optical storage has opened the door to a new era in image processing.

The Differences and Similarities for Archiving Images from Medicine, Document Storage, Satellite Imagery, Seismic Exploration, Astronomy and General Image Processing

Harold G. Rutherford, Ph.D.
Gould Inc., Imaging and Graphics Division
1870 Lundy Ave., San Jose, Ca. 95131

Abstract

This paper describes a family of image management systems which require massive archival capabilities. The Multimodal Acquisition, Archive and Review System needs a storage media which can be preserved for legislated time periods of five to ten years, with storage for childrens cases up to 21 years. The Multisensor Acquistion and Processing System (MAPS), which archives Landsat images and the Multimodality Image Management and Enhancement System (MIMES) for image management, processing and archiving require the ability to purge unwanted images and replace them with updated images. Seismic exploration image slice management and astronomic image management also require purging. As a result, the functionality will be dealt with briefly to demonstrate the broad application of the archiving media. The acquisition of images and the utilization of the end result of the processing are differentiating. The archiving needs they hold in common, as well as the subtle differences, will be emphasized. These aspects are: interfaces to network busses and erasable/non-erasable media needs. The role of interfacing to networks which are highspeed (13-28 Megabytes), those which have standard protocols and moderate speed (Ethernet) and those which support telecommunication will be presented. Data rates and the mediation of double buffering on the interface are addressed. Known error rates for the optical digital data disk and the error correction to the required accuracy for the application will also be discussed.

Introduction

The placement of this paper in this session and its title draw attention to the main purpose of the paper: Point out the needs for image storage and database management of the active and archived images in a wide spectrum of imaging modalities. While the list of disciplines being addressed is not complete, a major segment is being touched upon. Not included are such disciplines as image and patterns for robotics and CAD/CAM generated images. These, however, have similar needs.

This paper describes a family of image management systems which require massive archival capabilities. The Multimodality Acquisition, Archive and Review System needs a storage media which can be preserved for legislated time periods of five to ten years, with storage for childrens cases up to 21 years. The Multisensor Acquistion and Processing System (MAPS), which archives Landsat images and the Multimodality Image Management and Enhancement System (MIMES) for image management, processing and archiving require the ability to purge unwanted images and replace them with updated images.

Seismic exploration image slice management and astronomic image management also require purging. In each of these disciplines, a form of measurement of the amount and size of image storage required for image acquisition, viewing, and archiving will be given.

The major problems common to all such disciplines are 1) acquisiton of images, 2) distribution of images, 3) archive and recall of images from archive. The Multimodality Acquisition and Review System , a digital radiographic image management system includes components for

1. database management,
2. image access menus,
3. image display,
4. image enhancement,
5. networks, and
6. workstations.

This is used as an example of the setting in which optical disks and other high volume storage media play a major role.

History

The Multimodality Acquisition and Review System, the GS1000, was reported on in February, 1984 at the SPIE medicine XII in San Diego, USA [1]. The design considerations were reported at SCAMC, October 1983 [2]. The GS1000 was first shown in November, 1983 at RSNA, Chicago, USA. The digital lightbox and the digital alternator were the first components described in

May 1983 at PACS II in Kansas City [3]. Enhancement and image processing components were described in St. Louis at the ACR confrence on Computers in Radiology, May 1984 [4]. More sophisticated and general image processing algorithms are included in the Medical Image Processing Software (MIPS) to perform advanced work station processing. Many of these enhancement processes were suggested for inclusion in a PACS system, [5]. The PICNET component will be presented in Arlington, VA., USA [6]. It will be described briefly here since it is the network which connects the archive optical digital disks to the image database system. PICNET incorporates the realtime digital disk (RTDD) and its controller into a highspeed network.

Archive requirements for image management system

The functional use of the imagery in digital form may be used to demonstrate the broad application of the archiving media. While images generated by medicine and astronomy are distributed on thin film, space and satellite data are distributed on magnetic tape. The original data is digital and is most effectively and accurately worked with in that form. The generation and acquisition of images as well as the utilization of the end result of the image processing are different from one discipline to another. Acquisition is, however, possible digitally in each modality with the resulting implication on the amount of digital media required. The effects of image enhancement and pseudo coloring of medical imagery, of pseudo coloring of galaxies in astronomical imagery, of oil strikes in seismic imagery and enhancement and pseudo coloring of Landsat imagery is clearly seen. The archiving needs they hold in common are interfaces to network busses and erasable/non-erasable media needs.

Interfacing to networks which are highspeed (13-28 Megabytes), those with standard protocols and moderate speed (Ethernet, Hyperchannel) and those which support telecommunication each serve a need. Continual viewing by more than one user at a time requires the highspeed data rates from networks not on the host computer data bus. The PICNET of the Realtime Digital Disk by Gould is an example. Acquisition of digital data is able to use moderate speed star configured networks in most cases. Examples such as digital cardiology are exceptions which require highspeed due to the realtime acquisition rates by the radiographic modality. Archiving can make excellent use of the highspeed data rates and the facility for double buffering on the PICNET interface.

Known error rates for the optical digital data disk and the error correction to the required accuracy for the application are prime concerns to all at this conference. The cost of logic to provide error correction to an acceptable level increases the cost of the media overall.

Radiological

The quantity of digital radiographic images for a 540 bed hospital was reported by S. Dwyer III at PACS II in May 1983. Of 117,000 exams in a year, 29,367 were digital with a projected growth rate of 8% per year. Body and head CT images generate 600 images per day using 121 megabytes of storage per day. Nuclear medicine (NM) produces 310 images per day with 10 megabytes of storage. Ultrasound imaging (US) generates 740 images using 195 megabytes per day. Digital subtraction angiography (DSA) result in 155 images and 95 megabytes per day. Magnetic Resonance Imaging (MRI, also known as NMR) estimate at 2600 images and 162 megabytes. Digital chest would generate 100 images and 104 megabytes of data per day per machine. The total storage per day estimates at 687 megabytes per day, a large disk worth of data at todays estimates.

Astronomy

When a member nation served by the European Southern Observatory receives the product of ESO which maps the southern hemisphere, the package is 2 meters tall by 0.5 meters on a side containing thin film sheets of images which were digital images stored in floating point but are converted to thin film with the significant loss of ability to calculate. To preserve these images on disk in digital form would be a massive volume.

Space Missions

K. Castleman reports on the Jet Propulsion Lab's (JPL) experience with images received from space in several highlight years of space exploration. 4316 images were transmitted in 1964, and 7137 images in 1965 on the Ranger VI and VIII missions. The early Mariner Surveyor and later Mariner missions of 1973 and 1977 transmitted 86,000 images with 200x200 pixel accuracy and 16,000 images with 700x832 pixel accuracy. The total volume of this data base is just under 12 gigabytes.

Landsat

Landsat 1, 2, and 3 since 1972 generate four bands of 3000x2000 imagery every 20 seconds resulting in about one megabyte per second. Landsat 4 operating since 1982 generates in the neighborhood of 5.6 megabytes per second while functioning.

Seismic

Seismic imagery begins with Landsat imagery. This imagery is used to narrow the region for seismic explorations to a few square miles. The images which are reconstructed from seismic exploration are layers of 1024 squared imagery allowing three dimensional volume considerations.

Hardware components for IMS

The hadware components required for a fully functional Image Management System (IMS) touch the main categories of 1) acquisition, 2) display and enhance and 3) archive. Behind the scenes of these three major categories are system and network components. The subsections which follow identify the major components of Gould's GS1000 Multimodality Acquisition and Review System, one of the family of Image Management Systems.

Acquisition

For video acquisition, a frame synchronizer and a video switcher are used for multiple video inputs. An 8-bit digitizer provides digitized input to the system through the FD5000 Gould image processor and display unit for 525 line RS170 video images.

Software interfaces are available for the Ikonics CCD camera, the Optronics drum scanner and various vidicon cameras such as Cohu and Sierra Scientific for film digitization. The digitization may be run at highspeed, 15 to 30 frames per second, and is needed for applications such as heart catheterization labs. The Gould Real Time Digital disk is used to save acquired images at these rates.

A limited number of interfaces to radiology modalities are under development by Gould. These include serial and parallel interfaces as well as interfaces to the highspeed port of the RTDD controller. The highspeed port provides a high speed digital interface to digital modalities which acquire up to 30 images per second.

Digital alternator

The components for a Digital Alternator are included in Figure 1. This consists of 1) a Gould IP8500 high resolution image processor, 2) one or more 20 parallel head disks on a RTDD controller with burst rate of up to 26 megabytes/second, 3) three or four 1024 square high resolution monitors, and 4) a touch sensitive screen terminal. The digital alternator is a peripheral which displays and then enhances images on its own parallel bus. Response times are not appreciably affected by multiple digital alternators needed for a large volume of image handling. The only use of the host computer bus is for accesses to the database and the initial request to the IP8500 to display an image. All image data is on the PICNET parallel bus. PICNET and the hardware components of the IP8500 will be presented in more detail later and in [4] and [6] respectively.

Digital lightbox

It is the software design which distinguishes The Digital Lightbox from a Digital Alternator. The fundamental difference in concept is to display a series of images or single images of a patient in contrast to series of images grouped by modality such as Magnetic Resonance imaging or by a series of images for which a radiologist is responsible. Since differences are software, both concepts are supported on the same hardware. Only the different concepts of use distinguish between them. The next major section on software components will further differentiate the two concepts.

Archive

The archive segment of hardware is the most diversified. This is due to the wide variety of media, cost of storage per image and recall time per image. Among different renditions of the same media, the rate of recall and the cost of archive vary. Figure 2 shows some of the components, such as 1) cartridge tape, 2) magnetic tape, 3) disk drives, 4) and optical digital disks.

The important feature, for a large system, is the interface to a highspeed port of PICNET. The interface of an optical digital disk controller to PICNET provides a port to and from the host system where images may be sent or recalled one at a time. This provides for 1) no image data transfer on the system bus, avoiding degrading the system response to users, 2) high data rate transfers, 33 milliseconds for a 512 square image, and 3) insolation of PICNET from the slower data rates of the optical digital disk.

When permanent storage is required, optical digital disks are the most cost effective balance of storage cost and rate of recall. This applies to all fields of IMS. Where a large amount of archive is of temporary images rather than permanent images, a write once read only archive media is not satisfactory. Medical, Landsat, seismic and general image processing generate many intermediate images which, after a time, are in need of archiving permanently. However, the volume of intermediate images requires an interum archive.

Workstation

The basic components of a work station are those shown in Figure 1 for the digital alternator or lightbox. Primary differences are 1) a choice of 1024 or 512 square resolution for image display and processing, 2) configurations of monitors, 3) RTDD or conventional disk and 4) need for a touch sensitive screen terminal.

Three classes of monitor configurations are a) all color - from one to four in 512 and one or two in 1024 resolution, b) all monochrome - from one to six or even twelve in 512 and one to four for 1024 resolution, and c) a combination of monochrome and color, to examine individual images and, through color and pseudo color, examine similar images overlayed. Landsats and cardiology have this need commonly, while all IMSs could make use of it.

System

Presently the Gould system is based on the VAX family of host computers, using DSM as a layered language. Within the next twelve months when the potential of the microvax is realized the system may be converted to run on microvaxs. The components for the system are 1) the host computer, 2) one or more disks for the system and the database, 3) backup facilities and 4) printers and terminals.

Networks for image and demographic data

Networks for image movement fall into three general categories: 1) image acquisition, 2) image dissemination throughout the system and 3) transmission out of the system. A star video network is used for video image acquisition from external modalities. The result of video acquisition is a digital image. A star digital network is used for acquisition of digital images from external modalities. While video is coax cable or multibands of a broadband network for longer distances with minimal quality loss, Ethernet and parallel cables are used for digital. Of course, hand carried media, such as tape and floppy disk, serve just as well as current film transport mechanisms. This is viable until digital interfaces are installed and all images within the system are digital. Display back to the user is over coax cables to nearby video monitors and by multibands of broadband network for more distant video monitor displays.

With the exception of telecommunication of digital imagery to satellite medical facilities, there is no need to transmit digital images from the system to a remote viewing area or work station. The facilities for image display and image processing may all be kept within the range of PICNET and accessed via commands from a computer terminal and viewed on video monitors. Local image processing stations and disk storage is less cost effective than centralized image processing in the confines of distances spanned by PICNET.

Demographic data and data from word processing of physicians reports uses the system bus with computer terminal access from anywhere within the facility. Access can also be made from satellite medical facilities, physicians offices or homes using modems.

Software components for IMS

Images are the primary data of an IMS. However, a reliable database and software to add, delete and modify the database to provide accurate access to the images make the IMS functional. The design of the hardware and software to provide quick, lucid response to queries and fast, flexible display of requested images make it the IMS of choice. In the following paragraphs of this section we will highlight features in the design of the software to 1) acquire via video and digitally, 2) organize and display by means of the digital alternator and lightbox, 3) archive software to file and recall images and some 4) system software to provide access, security, image movement, and database integrity.

Automated acquisition

As mentioned previously, work is in progress to acquire images both via video and digitally from several radiographic imaging modalities. In the other disciplines mentioned, acquisition is physically in place. For each source of images a configuration file is maintained to provide parameters on image size, number of bits, rate of acquisition and other parameters to make acquisition as automated as possible. This individual treatment is necessary to support the many different manufacturers, modalities and models of radiographic equipment, in particular, and image generating equipment, in general.

Digital alternator and digital lightbox

The concept of a digital lightbox is a collection of hardware to serve as a repository, to display and to visualize digital images. In particular, 1) RTDD or conventional disk, 2) IP8500 image processor, 3) a collection of video monitors upon which to display images, and 4) software to display images from disk.

Just as a lightbox is a generic term for a backlit box upon which to view radiographs, the digital lightbox displays images on video monitors. A digital alternator is one or a collection of digital lightboxes with a particular functional organization implied. An alternator is used in radiology as a respository for current (often today's) radiographic films organized by modality, (such as DSA) radiological subdepartment, (such as neurology) physiological subgroup, (such as genitourinary). A digital alternator has all of these capabilities and more, all within the same physical viewing region. It is software and indirect list access to the image database which distinguishes one type of digital alternator from another. A digital alternator has access to all images stored on-line digitally.

In addition to the listed groupings, images may be grouped by radiologist, disease, anatomical organ, combinations of two or more such groups and groups of complete patient workups, all with shared access to images rather than a radiographic film physically being on only one alternator.

Archive - filing and recall

Once a series of images from an exam have been reviewed, the report generated and images marked for save or delete, the images to be saved are filed. Filing in an IMS emulates at least two of the radiology departments file rooms. The first is intermediate storage where attending, and referring physicians may view their patients images; access provided from surgery, intensive care units and physicians offices. The second is long term archive when a patients files are no longer active. The first is a digital lightbox with a less speed, less expense and greater storage volume, but with the same capabilities for digital alternator access. The second is a digital interface to the 26 megabyte burst PICNET through the highspeed port to an optical digital disk system.

A need arises here for a third level of archive and for multiple read and write capabilities of optical digital disk media. Long term archive could economically utilize optical digital tape when images can be transferred from read/write optical digital disk media. Recall is a process also referred to as order ahead, when images and exams are recalled from previous visits for reference when viewing current studies. Recall is basically movement of a copy of images from an off-line media to an on-line media, or from a slower media to a faster media, both on-line.

System - access/security/image movement control/database access and integrity

It is important to acknowledge that these facets of the system design are and will be adequately treated. Numerous other system considerations are also addressed. The details, while important, are not of general interest to regular users of the system and therefore will not be further elaborated on in this paper.

Archiving for image management systems

The diverse hardware available for archive is described in the section on hardware and pictured in Figure 2. Software interaction with the hardware has just been described in the software components section. Due to the current state of change in the laser optical media development and the absence of a variety of off-the-shelf products, archive will likely be a process of adding new cost effective media to an IMS in time to keep pace with growing image database needs.

Conclusions

Early in this paper we drew from several disciplines to give examples of the volume of image data to archive. We then presented an example of a working IMS for radiology. Finally, we stressed some important design criteria for archive the most likely media for archive, as we see it, in the immediate future. In talking with radiologists, there is some comfort in using a media which is not erasable. Perhaps the disk surfaces could be resurfaced for a nominal charge when image data has been transferred to more cost effective digital optical tape for long term archive.

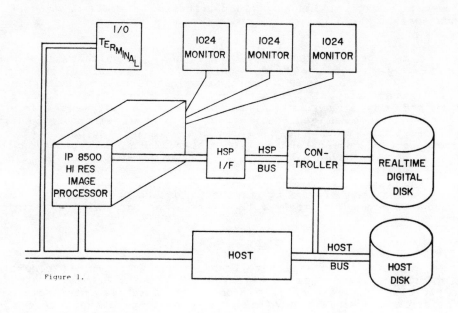

Figure 1.

PICNET CASE 5

S—SMD INTERFACE
U—UNIBUS PORT
A—A HIGHSPEED PORT
B—B HIGHSPEED PORT
HSP—HIGHSPEED PORT

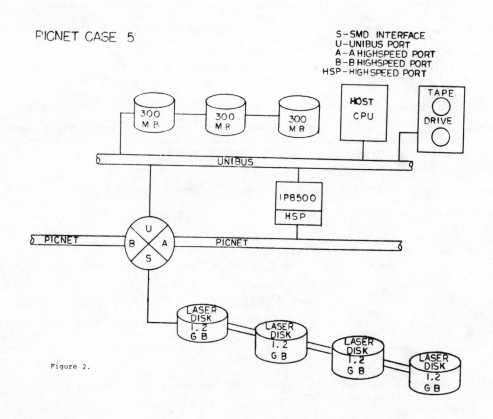

Figure 2.

Acknowledgements

The author acknowledges the support of N. Rutherford.

References

1. Rutherford, H.G., Gray, M., Hovey, J., Zima, P.J., Ogan, J.J., A Highspeed Multimodality Digital Image Management System. Medicine XII. 1984.

2. Gray, M., Rutherford, H., Efficacy of a Real-Time Digital Lightbox. SCAMS, Oct. 1983.

3. Rutherford, H.G., Gray, M.J., Digital lightbox, one of the integral pieces of PACS. SPIE Vol. 418, pp. 54. 1983.

4. Zima, P.J., Rutherford, H.G., Gray, M., Preliminary Results on Artifacts from the Image Enhancement of Nonfaithfully Compressed Images. A.C.R., 1984.

5. Rutherford, H.G., Reese, A., Gray, M., Zima, P.J., The Role of an Image Processing Realtime Digital Disk in a PACS system. Medpacs, 1984.

6. Rutherford, H., Patel, S., Raven, O., Colley, T., Gray, M., Implementing Image Enhancement Algorithms in High Resolution. A.C.R., 1984.

APPLICATIONS OF OPTICAL DIGITAL DATA DISK STORAGE SYSTEMS

Volume 490

Session 3

Standardization

Chairman
J. A. Th. Verhoeven
Philips Research Laboratories, The Netherlands

Joseph Zajaczkowski
Cherokee Data Systems, USA

FUTURE STANDARDIZATION DEVELOPMENT PROJECTS
WITHIIN THE AMERICAN NATIONAL STANDARDS ACTIVITIES

Joseph Zajaczkowski
Cherokee Data Systems
2334 Broadway Suite 206
Boulder, CO 80302

ABSTRACT

The Optical Digital Data Disk Study Group was established by the Standards Planning and Requirements Committee (SPARC) of the American National Standards Committee (ANSC) X3, Information Processing Systems in November of 1982. The study group was charged with identifying the standards necessary to accomplish data interchange on optical digital data disk. This work was completed in February 1984. As a result of the study group work a new X3 technical committee ANSC/X3B11, Digital Optical Disk was formed and held its organizational meeting in April 1984. The approved projects and proposed projects that will be under development and discussion by the committee are as follows: The approved projects are those concerned with unrecorded media standards for 120mm, 130mm, 200mm, 300mm and 356mm. The proposed standards projects for the committee include the recorded media standards that will complement the unrecorded media standards. Currently the committee is studying standards requirements in the area of labels and formats.

INTRODUCTION

Digital recording on optical media using laser technology has been under investigation and development by a number of laboratories and corporation for over a decade. Until recently, the activities projected promising results but did not warrant serious consideration by standards generating bodies. It is only in the last two years that serious consideration has been given to developing standards in optical digital data disk. This standards activity is a result of the technological gains in the industry and the importance the industry has ascribed to standards.

Historically, approval for project proposals within the standards generating bodies in the U.S. required that the project proposal identify a commercial activity that was already making use of a technique or a product that would be the subject of the standards proposal. The manufacture and marketing of such a product was a necessary fact before the standards new work proposal would be considered as such. The rate, of successful change of Optical Digital Data Disk (OD3) technology and the projection that by the year 1990, 39% of all information on line on rotating memory devices, 2000 trillion bytes, will be read optically rather than magnetically prompted the standards bodies to seriously consider developing the appropriate interchange standards.

STANDARDS DEVELOPMENT

The vast amounts of data that will be committed to be stored using OD3 technology highlighted the importance of considering all the physical and performance characteristics that would require and enable a user to successfully interchange data. With this in mind, the Standards Planning and Requirements Committee (SPARC) of X3, reviewed standards project proposals using OD3 technology and commissioned an OD3 Study Group in November of 1982. The Study Group under the chairmanship of W. M. Deese (1), was charged with identifying the standards that would be necessary in order to facilitate successful data interchange using OD3 technology Specifically, the study group was changed as follows:

1. Generate a scope and program of work
2. Identify areas requiring standardization
3. Generate the appropriate project proposals
4. Determine the need for a new technical
Committee

Interest in this work was demonstrated by the fact that there were more than 50 participants in the study group. The study group members represented media manufacturers, subsystem and device manufacturers, system integrators, and most important of all, users.

The first meeting was held in January of 1983 and the seventh and last meeting was held in February of 1984. The main issues considered by the Study Group were identified as:

1. The physical parameters of the media, i.e. the overall dimensions that allow for cartridge/drive interface requirements.

2. The recorded requirements which can include the recording code, servo information ECC etc.

3. File labeling and format consideration

4. Subsystem interface requirements.

The above issues are by no means definitive, and for the purposes of this report, they are listed as "generic" considerations. Within each topic the definition of those terms and the terms to be considered have been debated and are presently being compiled in a Vocabulary Document.

In order to facilitate progress on the main issues, working group within the study group were established. The working group considered 1) Recorded Media 2) Unrecorded Media 3) Vocabulary and 4) Project Proposals. These groups have been carried over into the Technical Committee X3B11, Digital Optical Disk.

It has to be kept in mind, that within the U.S. standards development groups on media, only those characteristics and features of the media (recorded and unrecorded) which constitute the minimum set of requirements that are necessary for data interchange be considered. Suffice it to say, performance requirements must be accompanied by appropriate test methods.

Technical Committee X3B11 commenced work on five approved project proposals in April of 1984 which are listed as follows:

1. X3 Project #0407-D: Unrecorded Optical Media Unit for Digital Information Interchange, Nominal 200mm (89.00 inch) Diameter.

2. X3 Project #0408-D: Unrecorded Optical Media for Digital Information Interchange, Nominal 300mm (12.00 inch) Diameter.

3. X3 Project #0409-D: Unrecorded Optical Media Unit for Digital Information Interchange, Nominal 120mm (4.72 inch) Diameter.

4. X3 Project #0456-D: Unrecorded Optical Media Unit for Digital Information Interchange, Nominal 356mm (14.00 inch) Diameter.

5. At the time of its final meeting last February the Study Group had also begun work to prepare several proposals for recorded standards projects.

In addition to the above, five proposals for Recorded Standards which complement the above were submitted to SPARC and should be considered for approval in July. The proposals bear the following Study Group document numbers:

1. 84-019R - Nominal 120 mm (4.72 in) diameter
2. 84-020R - Nominal 130 mm (5.25 in) diameter
3. 84-024 - Nominal 200 mm (8.0 in) diameter
4. 84-025 - Nominal 300 mm (12.0 in) diameter
5. 84-026 - Nominal 356 mm (14.0 in) diameter

Also to be considered by SPARC is a standards project proposal to address Labeling and Format considerations. The need for I/O interface standards requirements has been communicated to Technical Committee X3T9 and this group has already incorporated OD3 interface requirements in their work.

To date, the meeting of TC X3B11 have been well attended and the participants have been active. Liason with ECMA and Japanese standards groups have been established so that the activity in OD3 standards development in consistent if this spirit of cooperation is maintained, then the time lapse in getting the various standards published will be minimized and the differences, if any, will be academic.

REFERENCES

(1) W. M. Deese "The Need for and Persuit of Digital Optical Data Interchange Standards". Sixth Annual IEEE Symposium on Mass Storage, Vail, CO. June 1984.

Standardization Activities for Optical Digital Data Disk in Japan

Yoshinobu Mitsuhashi

Optoelectronics Section, Electrotechnical Laboratory
Tsukuba Science City, Ibaraki 305, Japan

Abstract

This note describes optical digital data disk (OD^3) standardization activities in Japan. We organized a study group for OD^3 in July 1983 at the Optoelectronic Industry and Technology Development Association. In June 1984 we formed a Technical Committee. In accordance with the Resolutions of the ISO/TC97 Advisory Group, we are preparing the organization as Secretariat of a new Sub-Committee SC23 : OD^3. The activities and status of these standardization programs will be presented.

Introduction

In the optoelecrronic industry, optical disks are considered one of the most promising technologies following optical communication. In recent years, technical development of the optical disk has been startling, with the marketing of the video disk and the digital audio disk followed by the document file and the code data disks, all of which illustrate its increasing variery. In 1982 we made an investigative report on the domestic and foreign technologies related to the optical disk as an activity of the Optoelectronic Industry and Technology Development Association (OITDA). This association was established in 1980 under the supervision of the Ministry of International Trade and Industry (MITI), with the aim of promoting research and development in optoelectronic technology and industry. In July 1983 we organized a Study Group towards the standardization of OD^3, stimulated by the Optical Storage Media Guidelines proposed by Dr. D. Chen's group at the Topical Meeting on Optical Data Storage.[1]

In 1984, with the increasing activity of the Study Group, A Technical Committee was established to prepare a draft of OD^3 Japanese Industrial Standard. The activities of these two groups aim at furthering OD^3 technology and increasing applications for the benefits of both users and manufacturers. Also, they will contribute to the establishment of international standards.

Study Group Activities

Between July 1983 and March 1984 we had seven times meetings with 33 companies participating. We provided a forum for discussion among present technologies, regarding the following items,

1. Problems towards the standardization of OD^3,
2. Optical storage media ; current status of several write-once materials, terminology of sensitivity, lifetime, noise, etc., and how to measure them,
3. Substrate materials ; glass, polymethylmethacryate, polycarbonate, aluminum : retardation, environment proof, flatness, shock damage, groove forming, and applications,
4. Optical head ; parts : high power laser diodes, laser noise, laser array, lens ; head : signal detection, drive mechanism, magneto-optics, phase-change; user problems,
5. Bit error ratio ; DAD, magnetic disk, video disk, system, media, add-on disk, erasable disk,
6. Access time ; method and mechanism, definition,
7. Applications ; features of optical disks : removable and transportable ; practical applications ; future applications : patent office, printing, medical information, sound information ; general requirements : environment, lifetime.

The Study Group proceedings, in Japanese, which consist of about 200 pages may be ordered from OITDA; 20th Mori Bldg., 7-4, Nishi-Shinbashi 2-chome, Minatoku, Tokyo 105, Japan, for ¥5,000 per copy.

In the fiscal year of 1984, the Study Group has been reorganized by OITDA as of this June and there are now 58 participating companies. Fig. 1 shows the classifications of the participating companies, and their activities. It should be noted that there are many companies classified as other than electronics ones.

This group aims to achieve information exchange regarding OD³ technology as well as exchange between users and manufacturers and will have a total of six meetings in this year. The items for discussion are new DRAW materials, erasable materials, pickup, high density recording, high speed bit transfer rate, history of an optical disk and applications.

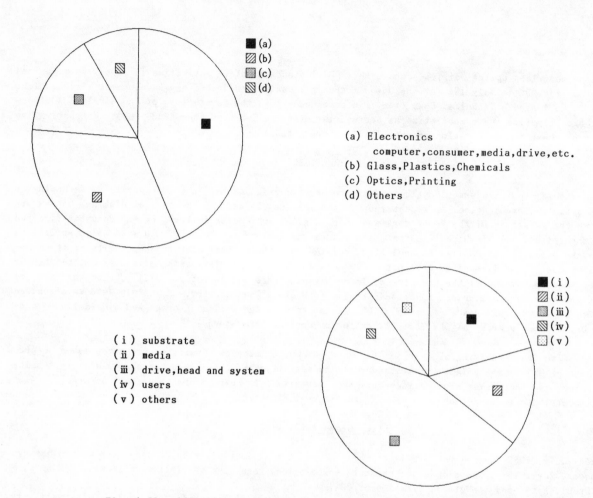

(a) Electronics
      computer,consumer,media,drive,etc.
(b) Glass,Plastics,Chemicals
(c) Optics,Printing
(d) Others

(i) substrate
(ii) media
(iii) drive,head and system
(iv) users
(v) others

Fig. 1 Classification of companies participating in the OD³ Study Group

OD³ Technical Committee

In June 1984 OD³ the Technical Commnittee (Dr. Y. Mitsuhashi, Chairman) was started at OITDA, for which a small number of members have been selected partly to avoid difficult situation due to too many opinions. The members are as follows,

Electrotechnical Laboratory,	Machinery and Information Industries Bureau of MITI
Standard Department of MITI,	Musashino Electronic Communication Laboratory
Electronic Industries Association of Japan,	Japan Electronic Industries Development Association
Japan Business Machine Makers Association,	Mitsubishi Rayon Co., Ltd.
Fujitu Laboratories,Ltd.,	NEC Corporation
Mitsubishi Electric Corporation,	Cannon Inc.
Olympus Optical Co., Ltd.,	Pioneer Electronic Corporation
Hitachi, Ltd.,	Sharp Corporation
Nippon Sheet Glass Co., Ltd.,	Oki Electric Industry Co., Ltd.,
Toshiba Corporation,	Sony Corporation
Matsushita Electric Industrial Co., Ltd.	

In the fiscal year of 1984, the OD³ Technical Committee will aim to make a report pertaining not only to the standarization of OD³ but also to wider aspects of optical disk technology. There are three working groups ; (1) the international liasion group for OD³ standardization with ANSI and ECMA, where we will study formalism and technology, (2) the technology investigating group where we will survey items such as terminology, technology, markets and patents of optical disks including the video disk, compact disk and CD-ROM, and (3) the competitive technology investigating group. In the fiscal year of 1985, the OD³ Technical Committee will concentrate it efforts to make standardization specifications.

We would like to note that before beginning our efforts of OD³ standardization we need to understand more broadly the still rapidly improving optical disk technology.

The tentative schedule of the OD³ Technical Committee

FISCAL YEAR	1984	1985
PLAN MAKING		
RESEARCH OF TECHNOLOGY AT DOMESTIC & OVERSEAS		
RESEARCH OF STANDARDIZATION AT DOMESTIC & OVERSEAS		
TERMINOLOGY		
STRUCTURE OF SPECIFICATIONS		
PROTOTYPE SPECIFICATIONS		

ISO/TC97 SC23 OD³

Recently the ISO/TC 97 Advisory Group has approved the formation of a Subcommittee SC23 on optical digital data disks and the Japanese Member Body has been appointed as Secretariat of SC23.[2]

We have accepted this resolution with much pleasure and are now preparing the organization.

The first SC23 meeting will be held in Tokyo from May 29-31,1985. Dr. Junichi Shimada, a Director of the Radio- and Opto- Electronics Division of Electrotechnical Laboratory, will serve as a chairman of this meeting. General affairs such as financial services, receiving and sending letters and documents from and to other countries will be performed by the Japanese Standard Association; 1-24, Akasaka 4-chome, Minato-ku, Tokyo 107, JAPAN.

Acknowledgements

The author thanks Messrs. T. Honda, K. Fujiwara and Y. Yamaya for their help with OD³ standardization activities of OITDA and Dr. Di Chen of OPL, Mr. M. Deese of STC and Mr. D. Hekimi of ECMA for providing information relating OD³ standardizatin, ANSI and ECMA activities, respectively.

References

1. Olson,T., "Optical Storage Media Guidelines", Technical Digest of Optical Data Storage, TuB1, January 17-20, 1983, Incline Village, Nevada
2. Resolution 10 (Rev), Resolutions of ISO/TC 97 Advisory Group, 1984-05-16

Standardization activities for optical digital data
technology in European Computer Manufacturers
Association (ECMA)

Verhoeven, J.A.Th.

Optical Storage International
P.O. Box 218, 5600 MD Eindhoven, The Netherlands

Abstract

Optical storage of data as a technology has been developed for many years. In the very near
future optical storage products will be available in the market place. Optical disk systems
vary substantially in disk diameter, recording method, recording material and production
technology. Standardization will be necessary if such systems are to be brought into wide
use through interchangeability.
Standardization in this field was started within ECMA (European Computer Manufacturers
Association) [1] in 1983. Within ECMA - Technical Committee TC16 (Rigid Magnetic Disks)
a Task Group was formed to investigate the need for OD3 standardization.
The findings of this study group led to the start of a new Technical Committee (TC31) on
Optical Digital Data Disk Standardization. This TC31 has started recently. The findings
of this Task Group in terms of the importance of standards for media interchange, the
categories of standards needed, the programme of work and the liaisons with other standardization organizations will be reviewed.

Introduction

Optical storage of data as a technology has a history of many years. Questions with
respect to stability, archivability (shelf life), data integrity, and producibility of the
media themselves were, until recently, the reasons for a waiting attitude. Recent
significant advances with the optical media have brought optical digital data disk to the
commercial data processing market place. A variety of American, European and Japanese
firms presented prototype optical-memory systems and media at computer and micro graphics
trade shows this year. Several companies have announced or are already busy with "initial
user tests" (beta testing). Commercially availability is at hand.
The first generation of optical digital data disks (OD3) will be based almost exclusively
on lasers writing data by distorting thin (metal) films. In some systems, the laser burns
holes in the thin (metal) film. In other processes blisters are araised by lasers. In
still others, a change in the index of reflectivity is created by the laser beam.
Regardless of which technique is used, the pattern of um-sized holes, blisters or marks
in the medium surface causes the readback laser beam to be deflected at specific intervals, thus reproducing the original bit pattern.
In addition, the removability of optical disks makes it possible to develop automatic
disk-changing mechanism and to use OD3 disks for data exchange.
These characteristics of the medium and the potential possibilities of optical recording
(contactless recording, easy drives, no dust-related head-crash problems) are the basis
for predictions that by 1990 a substantial part of non-changing digital and image data
will be stored on low-cost, removable, high-density, high data integrity, easy-to-handle,
long-life optical media.

Role of a standard and primary objectives of OD3 standards

The industry is clearly in need of a removable form of storage media which can hold very
large amounts of both digital data and documents in machine-readable form and which can be
accessed very rapidly. It is widely felt that the optical storage of digital data
presents both a near-term and a long-term solution to this need. The role of standards
for optical disk will be quite vital in the commercialization of removable, easily exchangeable optical disks for data and document storage and retrieval.
The primary objectives of the proposed standards should be:

i)   To provide a common base of understanding to allow manufacturers and users of this
      technology to achieve the cost and efficiency benefits which this new data tech-
      nology offers.

ii) To provide industry-acceptable standards and means of measurement of the various parameters that are encompassed in high-density information storage systems.

Derived from or related to these objectives a standard should, in addition:

- Reduce the proliferation of non-compatible media and cartridges.
- Improve and facilitate the second supplier (or multiple source) appearance.
- Underline the reliability of the new product.
- Achieve interchangeability of recorded disks between systems for distribution of data. The standardization should include duplicated disks too.
- Achieve low production costs.
- Facilitate the acceptance of the medium as a "legal" document.

Discussion within ECMA learned that the standardization has to be started now for several reasons:

- First generation of products are being settled now.
- Technology is at a stage where the concept can be commercialized.
- Products (both media and drives) are or will be announced this year.
- Large volumes of products will be shipped next year; substantial larger volumes in 1986.
- Now the products come to the market, patents will easier be disclosed.

Categories of standards

With respect to the question what will have to be standardized one realizes that the standardization scope can include both the unmounted disk as an OEM product, the disk with cartridge as a physical interchange medium, the recorded disk as an information interchange medium as well as the I/O system interface (drive-medium system as a black box). In ECMA standardization starts with the medium and cartridge. The standardization of this medium-cartridge system can be subdivided into several groups:

- Physical disk standards (dimensions of the disk, optical parameters, sensitivity for recording, dimensions of the laser induced marks, environmental parameters, etc.).
- Cartridge enclosure standards.
- Servo information format standards (grooved or not, modulation system, tracking system, etc.).
- Data format standards (labelling, error detection characters, data field, file management, etc.).

Although a similar subdivision may involve, for several disk concepts, some complications, it certainly represents a sequence of steps along which a standardization can be reached. In addition, within each group several versions of standards have to be considered. As far as (outer diameter) size of the disk is concerned, the standard will have to deal with several sizes (e.g. 356 mm, 300 mm, 200 mm, 130 mm and 120 mm). In the 3rd group one can distinguish grooved and non-grooved versions. This may lead to a standardization based upon a combination of (sub)standards belonging, respectively, to the four groups mentioned above. Whether or not this will happen is dependent on the relevance of the different combinations, on forces in the market and on activities of the standardization groups.

Scope and programme of work of ECMA TC31

Activities in ECMA to develop OD3 standards started in 1983. This work is now being done in ECMA technical Committee 31.
The scope of ECMA TC31 is defined as: to identify and develop the minimum number of standards necessary for data interchange by means of optical digital data disks (OD3).
The programme of work of TC31 yields:

1. To identify and develop the sets of standards needed to cover grooved and non-grooved disks.
2. To develop a standard for a cartridge common to both types of disks.
3. To develop, in liaison with TC15, a standard for file structure and labelling applicable to both types of disks.
4. To develop a standard for selected configurations for data interchange.
5. To monitor technological developments in the field of OD3.
6. To maintain liaison with other standard organizations in order to present ECMA proposals to them and to make comments to their proposals.

Quite important for standardization is a general accepted vocabulary. The glossary of terms as used in the past at different meetings (Guidelines [2], ANSC X3 (see below) [3], Workshop at the NBS [3]) will be incorporated in the TC31 formulations.

Several ordinary and associated members of ECMA (3M, O.S.I., Sony, S.T.C., Thomson) have already sent in proposals for OD3 standards. Several major European and non-European companies have shown interest in OD3 standardization activities in TC31. The frequency of meeting will be about 4 times a year. It is generally thought that the time window for coming to a standard will be a very narrow and critical one. The standardization of optical digital data disks has already appeared in the ISO-arena. It is planned to have a well prepared and well supported ECMA proposal available within two years from now.

The American National Standards Committee X3 (ANSC X3) had started the development of OD3 standards in 1982. This committee ANSC X3-B11 operates under the procedures of American National Standards Institute (ANSI) and a formal liaison has been established between this committee and ECMA TC31.

Within Japan the Electrotechnical Laboratory and Technology Development Association has begun formal work on OD3 standardization. Recently, the International Organization for Standardization ISO (ISO-TC97) has set up a sub-committee (SC23) in charge of standardization of OD3. Japan will act as secretariat of this ISO-TC97/SC23 sub-committee. It is expected that this sub-committee will initiate meetings between the standards development activities in Japan, ANSI and ECMA.

1   European Computer Manufacturers Association, Rue du Rhône 114, CH-1204 Geneva, Switzerland.

2   T. Olson, "Optical Storage Media Guidelines", Topical Meeting on Optical Data Storage, Incline Village, Nevada 1983.

3   Proceedings of the National Bureau of Standards/National Security Agency Workshop on Standardization Issues for Optical Digital Data Disk (OD3) Technology, held at National Bureau of Standards, Gaithersburg, Maryland, June 1-3, 1983, NBS Special Publication 500-111.

**The Institute for Computer Sciences and Technology
at the National Bureau of Standards (NBS/ICST)
optical digital data disk (OD^3) standardization activities**

Jean B. Freedman

National Bureau of Standards
Room A216, Building 225, Washington, D.C. 20234

Abstract

This presentation will describe optical digital data disk (OD^3) standardization activities including the NBS-sponsored Federal Council on Computer Storage Standards and Technology (FCCSSAT); the National Bureau of Standards/National Security Agency Workshop on Standardization Issues for OD^3 Technology; and the NBS/ICST participation in the voluntary OD^3 standards process.

All of these NBS/ICST activities provide a forum for discussion among current and potential OD^3 users and suppliers, regarding the prospects for OD^3 data interchange standardization. The status of these standardization programs will be presented, along with their objectives for determining the directions for the development of OD^3-related standards and guidelines for the U.S. government.

Introduction

The diversity of optical digital data disk (OD^3)* dimensional and optoelectrical characteristics and applications represent a wide variety of standards requirements. To assess these requirements and develop standards and guidelines for the range of OD^3 media, systems and their applications, the Institute for Computer Sciences and Technology at the National Bureau of Standards (NBS/ICST) is supporting standardization activities for Federal ADP (automatic data processing) managers and OD^3 media and systems developers in the private sector. A description of these activities follows.

The Federal Council on Computer Storage
Standards and Technology (FCCSSAT)

To provide an opportunity for Federal ADP users to influence the standards process, so national standards and Federal guidelines better reflect their needs, the NBS/ICST has formed a new "Federal Council on Computer Storage Standards and Technology" (FCCSSAT)—pronounced "fix-it."

The objective of the FCCSSAT includes the following:

. To provide a forum for discussion of timely information on OD^3-related standardization issues, concurrent with happenings in the national and international computer storage standards committees.
. To provide a forum for enhanced interaction among Federal agencies.
. To provide an opportunity for vendors to present technological advances to Federal users.
. To support Federal Government representation in computer storage standards writing committees.

Initially the scope of the FCCSSAT will be the development of guidelines for the selection, operation, and evaluation of OD^3 storage systems. Subsequently, other computer storage media and systems will be included.

FCCSSAT members will also have the opportunity to review and contribute to those OD^3 technology standards and guidelines being developed under the auspices of the American National Standards Institute (ANSI), in addition to Federal Information Processing Standards (FIPS), that are published by the U.S. National Bureau of Standards.

Computer-driven printing technology guideline

An assessment of publishing systems that incorporate OD^3 storage systems will be proposed by NBS/ICST as a guideline, to be developed, to provide assistance to the user in establishing quality requirements, evaluating vendor samples, and equipment performance through the evaluation of the total reproduction process of a system comprised of a scanner, OD^3 storage system, and laser printer or plotter. The development of a "test image" that will ultimately qualify the process, will be an integral part of this program.

This program includes an initial assessment of the acceptance/rejection criteria used to evaluate an installation's contrast, typography, resolution, tone reproduction characteristics, consistency, and word and character spacing requirements of originals and their reproductions.

---

*The terminology 'optical digital data disk (OD^3)' is gaining acceptance in the computer storage media community in order to differentiate this computer peripheral storage technology from optical disk technology which is used in audio and video home entertainment systems.

Test images will then be specified to represent the ranges of the previously established image quality criteria. The selected test images will then be scanned, stored (OD³), and printed. Note: The term "printed" used throughout this document refers to any type of imaging on paper or metal film.

Subsequently, production documents will then be scanned, stored (OD³), and printed. Output that is considered "acceptable" will then be correlated with specific values on the test image(s) to "calibrate" input and output of the system.

To further analyze the system's reproduction of the test images, the execution of a tone reproduction study will be recommended. Such a study includes the evaluation of how original copy changes at each step in the reproduction process. Specifically, the study documents how the original compares with its generations, including the final copy[1]. **See figure 1.**

From these analyses, the inherent deficiencies of the system can be determined and special techniques can be applied to compensate for the "gains and losses" in quality. From the data collected, predictions can be made about the final reproduction. The data collected also serves as information that both the user and equipment manufacturer can use to record the operating history of the system. It is recommended that the "test image" be stored (OD³) and also maintained in hard copy, so that comparative printouts, of both forms, can be made periodically, to document the scanner/storage medium/printer performance.

The extensiveness of this proposed guideline and the development of other guidelines for OD³ applications will reflect the priorities established by NBS and the FCCSSAT membership. Council meetings will be held in Washington, D.C., beginning September 1984. Meetings are open to all Federal employees. Individuals in the private sector may attend by invitation.

NBS will solicit contributions and requests for discussion topics and technical development projects for the FCCSSAT. Contact Jean B. Freedman, National Bureau of Standards, Room A216, Building 225, Washington, D.C. 20234, telephone: 301/921-3723.

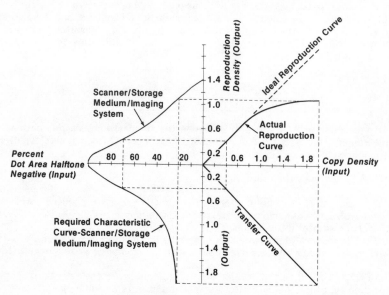

Figure 1- Tone Reproduction Study

This tone reproduction study represents "relative print quality characteristics" of a digital imaging system, where the maximum solid ink density (SID) is 1.1. Assume that all processes following the original scan are consistent, reproduction of the original is desired and the printed output represents the combined tonal responses of the scanner/storage medium, such as OD³, and imaging system.

<u>NBS/NSA workshop on standardization issues
for OD³ technology</u>

The NBS/NSA Workshop on Standardization Issues for OD³ Technology was held in June 1983 at the NBS in Gaithersburg, Maryland. The purpose of the workshop is to examine the emerging OD³ technologies and applications, discuss the prospects for standardization of OD³ media and systems, and to provide technical information and guidance towards the development of future standards, based on the expressed needs of users and suppliers.

Current and future OD³ users, manufacturers, and technical experts presented papers, and participated in both panels and open discussions at the workshop.

Forty-six current and potential optical disk users participated, representing twenty-seven different Federal and commercial organizations. These users were familiar with the applications of OD³ technology, were original equipment manufacturers (OEM), and/or represented system integration companies.

Twenty-seven OD^3 developers and suppliers participated; they represented sixteen organizations; four were consultants, one was a trade association delegate, and four were members of organizations engaged in computer storage media technologies other than OD^3.

Workshop topics

The workshop presentations and panel discussions were selected to define the characteristics of an optical media unit and the digital data channel interfaces as related to drive performance; the applications for digital optical storage; test methods for determining relative media and data life; legal and archival requirements of an optical media unit; and the potential directions for OD^3 data interchange standards.

Workshop summary

Many workshop participants noted that there were numerous complexities associated with the OD^3 data interchange parameters, such as media sizes, reflection and transmission characteristics, error detection and correction schemes, and numerous other parameters that are incorporated in the functional design of OD^3 media and systems.

The participants also noted the timeliness and importance of OD^3 data interchange standardization activities and gave priority to the development of an optical digital data disk lexicon. Many of the workshop participants also expressed an immediate need for standardized test and evaluation procedures for OD^3 media responses, life expectancies, and environmental requirements.

Availability of proceedings

A proceedings which consists of synopses of the workshop presentations and panel discussions is available from Superintendent of Documents, U.S. Government Printing Office, Washington, D.C. 20402, U.S.A., for $7 per copy (add 25 percent for other than U.S. mailing)[2]. Order by: SN 003-003-02573-9. Government Printing Office Order Desk: (202) 783-3238.

NBS/ICST OD^3 standards committee participation

Members of the NBS/ICST participate in the OD^3-related X3 Information Processing Systems Committees of the American National Standards Institute (ANSI). Jean Freedman and Tom Bagg are members of X3B11, Optical Digital Data Disk; James Upperman: X3L5, Data Formats, File Structure, Stored Data Organization, Labels and Descriptors; William Burr: X3T9, Interfaces for Mass Stoage Systems, including magnetic tape and disk and the optical digital data disk; John Little: X3L2, Codes and Character Sets; and Dennis Branstad: X3T1, Data Encryption.

The NBS research program for the archival life testing of OD^3

An NBS research program that covers the characterization and assessment of the life expectancy of optical media used for data and image storage, "NBS Research Program for the Archival Life Testing of Optical Digital Data Disk (OD^3)," is currently ongoing under the direction of Mr. Sidney B. Geller of the National Bureau of Standards[3,4]. A report on this program will be given at this SPIE conference. This NBS program is also published in The International Society for Optical Engineering (SPIE) Proceedings of "Applications of Optical Digital Data Disk Storage Systems," Brussels, Belgium, June 1984.

Conclusion

Through involvement in the FCCSSAT, NBS/NSA workshop, voluntary standards committees and the NBS Program for the Archival Life Testing of OD^3, NBS/ICST will continue to contribute to the development of OD^3 applications and technology standards and guidelines. NBS/ICST has always had a strong interest in state-of-the-art computer storage technologies, such as OD^3. In response to the growing interest in the prospects for OD^3 technology standardization, NBS/ICST will continue to research the related science and engineering developments, with an objective of developing user and industry-oriented guidelines.

References

1. Freedman, J. B. "Electronic computer printing technologies, measurement techniques, applications and management controls," Policy Studies Corporation (U.S.A.) lecture for Pappersgruppen AB and the magazine Pappermed Tryck, Stockholm, Sweden; October 1982. 234-239 pp.
2. Freedman, J. B. ed. Standardization issues for optical digital data disk (OD^3) technology. Nat. Bur. Stand. (U.S.) Special Publication 500-111; April 1984.
3. Geller, S. B. The National Bureau of Standards research program for the archival lifetime analysis of optical digital data disks (OD^3). International Society for Optical Engineering (SPIE), Brussels, Belgium; June 1984.
4. Park, J. R. Bibliography of the literature on optical storage technology. Nat. Bur. Stand. (U.S.) Special Publication 500-107; December 1983.

The National Bureau of Standards research program for the archival lifetime analysis of optical digital data disks (OD³)

Sidney B. Geller

National Bureau of Standards
Room A216, Building 225, Washington, D.C. 20234

Abstract

The Institute for Computer Sciences and Technology at the National Bureau of Standards (NBS/ICST) is embarking on a research program into the life expectancy properties of optical digital data disks (OD³). This paper discusses lifetime concepts in a general sense and some philosophies and objectives which will underlie the NBS/ICST archival program when it is initiated.

Whereas the associated OD³ systems including optical disk drives, servos, lasers, and optics are usually replaceable (providing that they do not fall into the one-of-a-kind category), data contents which are lost due to the OD³ media failure may be irretrievable. Therefore, the principal archival lifetime factors to be investigated by NBS are related primarily to the life expectancies of the OD³ media and media structures. The initial program efforts will be towards determining the system independent degradation and failure mechanisms of the media materials through static optical, chemical, and physical testing. Subsequently, where possible these static test results will be correlated with the results derived from dynamic tests of the system dependent degradation and failure mechanisms of the media. These program activities will serve as the background for the development of a quantitative OD³ lifetime prediction theory and an archival care and handling methodology.

Introduction

The Institute for Computer Sciences and Technology at the National Bureau of Standards (NBS/ICST) Computer Storage Media Group has been involved with the production of Standard Reference Materials for computer magnetic storage media since 1969[1]. In addition to this ongoing major function, the group has performed studies relative to the effects of various energies on the stored data[2,3] and has produced publications relative to the care and handling of these media on both a daily and long-term basis[4]. Group members participate in voluntary standards activities in organizations such as the American National Standards Institute (ANSI) and the International Standards Organization (ISO).

Presently, the Computer Storage Media Group is embarking on a program in the area of optical digital data disks (OD³) which includes both active standards committee participation and research into the life expectancy problems of these optical disk media. The possibilities for Standard Reference Materials in this new, rapidly growing technology will be dependent upon the activities in the OD³ marketplace and requests emanating from the voluntary standards community and industry[5]. This paper discusses lifetime concepts in a general sense and some philosophies and objectives which will underlie the NBS/ICST program when it is initiated. This program will be coordinated with a number of the ongoing life expectancy and material evaluation programs at NBS.

General life expectancy factors[6]

Lifetimes

The concepts of lifetimes are closely identified with those of reliability. Reliability is related to the probability that a system or device will perform its intended function for a specified time under specified conditions. In the case of lifetimes for which reliable operations extend for a few thousand hours, a statistical concept such as the mean time between failures (MTBF) is a valid unit of reliability measure. Unfortunately, the more reliable the device or medium under test is, the more difficult and costly it becomes to measure and determine its long-term reliability. If the tests on a highly reliable device are carried out in real time, then it is possible that the device will become obsolete due to technological advances by the time the results are completed. Therefore accelerated stress (aging) testing must be undertaken. However, the device or media response to an accelerated stress test does not always assure predictability of its response under normal stress environments, particularly in the case of media that display varying or nonlinear responses to different types of stress. The application of reliability theory is often valid in the case of accelerated testing when used with concepts such as the Arrhenius and Eyring models[7].

The fact that there are two different media decay rates, i.e., static and dynamic, tends to complicate the design of a valid life expectancy experiment. Static lifetime tests are usually related to the medium and data contents individually and are often performed on an accelerated basis. However, dynamic lifetime tests are more system oriented and are considerably more difficult to define, design, or evaluate quantitatively, because there are many more interacting variables. For completeness the term "lifetime" should include the following three areas: system lifetime, media lifetime, and content lifetime.

System lifetime. By "system" is meant the total grouping of all of the interrelated components such as the hardware, the storage medium, and the encoding/decoding schemes. System lifetime has:

. End-of-life due to obsolescence
. End-of-life due to the inability of the system to perform reliably beyond a certain time in a specific environment
. End-of-life due to loss of one-of-a-kind transducer or encoding/decoding keys

End-of-life due to obsolescence indicates that there is a point beyond which it is no longer feasible to maintain a system in view of technological and possibly sociological advances which may have occurred. Failure to detect this end-of-life through periodic cost analysis and operational studies leads to inefficient and costly operations.

End-of-life due to the inability of the system to perform reliably beyond a certain time in a specific environment may be difficult to analyze and to correct because it usually stems from both static and dynamic interacting causes which depend on system operation philosophies. This aspect of archival system lifetime should be considered through reliability studies at the very outset of the system design and development and not as an afterthought (consider the case of pure tellurium OD^3 recording surfaces). However, there is an optimum extent to which the system reliability should be developed, since after this further attempts at increasing the lifetime may result in increased system costs beyond a reasonable level.

Storage medium lifetime. This lifetime is the length of time for which a medium can perform its storage function over the entire range of required environmental and operating conditions. It is often difficult to design a dynamic decay experiment which will lead to valid judgments about the probable end-of-life of the storage medium because the dynamic test procedures can intrude their own characteristics into the results so that it is difficult to separate the cause and effect variables.

Considerable care must be used when attempting accelerated lifetime studies because in many instances accelerated lifetime testing procedures yield results that are not valid. As an example, by using higher temperatures, we may wish to produce some observable heat breakdown of a reel of magnetic tape in a reasonable length of time; however, in such a test we must be careful to remain below temperatures that can produce abnormal phase changes in the magnetic particulate material because heat effects are not cumulative for the tape. That is, no amount of heat below the critical value can produce a phase change in these magnetic particles no matter how long the tape is subjected to that lower temperature. Analogous heating effects on optical disk media should also be considered; for example, crystalline to amorphous changes.

The results of simulated test procedures must also be examined carefully for anomalous effects. For example, it has been found in a head-tape wear simulation test that the relative wear results among a group of magnetic computer tapes vary and even reverse themselves dependent upon the substitute metal used to simulate the true head materials.

Content lifetime. Content lifetime is the length of time for which the stored information is retrievable. It is possible for the contents to remain intact in storage but to become irretrievable due to the loss of the encoding/decoding keys or of one-of-a-kind transducers. As an example: the use of time-shared computer systems may increase the risk that unauthorized persons may gain access to confidential data banks. In order to protect these data, it may be necessary to encode them into a more cryptic form--this increases the probability of data inaccessibility through code losses. The loss of encoding/decoding keys has a finite probability of occurrence in a nuclear age. Redundant storage at diverse locations on different media is one method for protecting the contents.

NBS/ICST OD^3 lifetime program philosophy

Since the data lifetime or archivability of optical digital data disk OD^3 systems is principally a function of the storage media and media structures, this program will initially investigate the long-term stability of these components. The NBS/ICST OD^3 life expectancy program will begin with the non-erasable OD^3 types, and in view of the considerable ongoing research and development efforts by many organizations to achieve the first fully acceptable type of marketable and hopefully standardized optical digital data disk medium, the program must consider a variety of these media and media structures. Although there is considerable interest in the use of optical disks for image processing, the NBS program will be directed towards the use of these media primarily as computer data storage devices with data processing applications similar to computer magnetic tapes and disks. The quality requirements of this type of application are significantly more stringent than for video image processing and storage. For example, while a (corrected) bit error rate (BER) of approximately 10^{-7} is usually acceptable for video images, a BER of 10^{-12} to 10^{-14} is the objective for data processing and storage. The chosen types will be capable of write once, direct-read-after-write (DRAW) operations and be useful for both large mass storage applications and office environments.

An unresolved problem at the present time is the lack of a direct analytic relationship between changes in the system independent optical, chemical, and physical parameters of the OD^3, and the effect of these changes on the life expectancies of the OD^3 media and media structures. The data derived from the measured deterioration in these static parameters as a function of time and environment have not as yet led to a model or an in-depth discipline for lifetime predictability. That is, while there is a body of knowledge relative to the effects of impurities, flaws, and other imperfections on the intrinsic and extrinsic properties of the materials which make up optical disk structures, there is much to be learned about the complex problem of their contributions to device degradation and the eventual OD^3 failure with time. These failures are usually observed and measured during system operations through defined end-of-life factors such as excessive bit-error-rates, decreased signal-to-noise ratios, and decreased output signal amplitudes relative to some nominal value. It is proposed that the OD^3 media be investigated as entities with life expectancy (and failure) properties related to their static structures and viewed, where possible, separately from their system dependent failure mechanisms which are functions of the data configurations, the laser beam system, the optical system characteristics, the disk speeds, the servo systems, etc. Data deterioration and failure which is caused by excess read laser beam power is an example of a system dependent failure mechanism. Since media-system interaction is minimized in a long-term archive, more effective media maintenance in the archive is made possible through an understanding of static media qualities.

Inevitably, however, the final end-of-life will be determined through the inclusion of, and the measurement of, media interaction with the total operating systems.

It is hoped that the end-of-life extrapolations which are based upon accelerated stress testing can be augmented through the development of the aforementioned prediction discipline which is based on the system independent factors. This development will incorporate the pertinent resources that are available at NBS as well as the rigorous application of reliability theory, various lifetime testing methodologies, physics of failure, and failure analysis and diagnosis. Complex reliability measuring techniques such as "correspondence analysis" for failure analysis may be applied if the expertise is available to the Group[8].

A failure is characterized as a state in which the media or media structure does not fulfill their intended function. A failure or aggregate of failures may lead to end-of-life. Some of the following OD^3 failure tests will be considered:

1) Media failure due to the method of care and handling of the OD^3's in long- and short-term storage environments.

2) Media failure due to changes in the static structural characteristics of the OD^3 after being in both stressed (temperature, humidity, gas, vapor, etc.) and non-stressed environments both prior to and after recording and storage.

3) Interrelationships among failure types; e.g., the effect of chemical variations on OD^3 optical parameters, etc. It is possible that the degradation of a particular variable may not produce a failure mode. It is also possible, however, that the degradation of this particular variable may affect another variable which can then cause an actual failure.

4) Media failure due to exposure to external energy fields such as nuclear, x-ray, high voltage, etc. Interestingly, very preliminary tests in the ICST Computer Storage Media facility showed that electrostatic discharges (i.e., arcing) onto the surface of an optical medium produced visible and permanent tracks onto the medium surface. The operational effects of the discharges and tracks are not yet determined.

5) Media and/or structural failure due to the application of mechanical energies such as impact, pressures, torsional, as well as surface distortion due to scratches, etc.

6) Relative failure probabilities for the different physical approaches to data recording (e.g., pits, bubbles, etc.) and for different structural types.

7) Relative failure probabilities for different modulation methods.

Where applicable, some of these tests will be applied to samples drawn from the same manufacturers lots. Failure diagnosis and analysis will be instituted as the study is being structured and progresses.

The initial NBS efforts have consisted of a literature search for OD^3 papers whose titles and abstracts contained terms such as "archival," "lifetime," "life expectancy," and "stability." A number of papers have been read to date and entered into a special format which includes categories such as:

1) Type of medium and structure
2) Recording process
3) Tests performed
4) Test objectives
5) Test equipment or methods
6) Archival claims (by authors)
7) Comments on archival claims
8) Conclusions

Examples of "Tests performed" which were encountered to date in the qualifying papers are as follows:

1) Physical analysis (surface layer dimensions and uniformity, macro and microstructures, pits, bubbles, etc.)
2) Accelerated stress (aging) (Z/AD testing, various temperature-humidity environments, gas tests, etc.)
3) Optical analysis (small and large surface areas, reflectivity, transmittivity, absorptivity, etc.)
4) Chemical analysis (Secondary Ion Mass Spectroscopy (SIMS), Electron Spectroscopy for Chemical Analysis (ESCA), etc.)[9]
5) Recording surface sensitivity analysis (thermal analysis, including factors such as thermal diffusivity, etc.)
6) Signal analysis (signal-to-noise ratio (S/N), pilot-to-noise ratio (P/N), carrier-to-noise ratio (C/N), bit-error-rate (BER), harmonics, etc.). Note that "signal analysis" is a dynamic system test.

A list of the instrumentation used for the chemical, optical, physical, and signal analysis is also being compiled from these formats, and a summary describing each type is being written. Their availability in NBS or other laboratories or the potential purchase of these instruments for the ICST Computer Storage Media Group OD^3 life expectancy program is also being investigated.

It is suggested that the following two measurement techniques which were not found in the OD^3 literature to date be considered: laser spallation[10] and Mossbauer spectroscopy. In particular, laser spallation which is a technique for detaching thin films from substrates by impinging a pulsed, high energy laser beam onto the substrate can yield information about threshold values for film-substrate detachment. The feasibility of the spallation process will require considerable investigation.

Available scientific disciplines and resources at NBS include surface chemistry, optical analysis, material sciences, thin film deposition and measurements, and engineering technologies including statistical engineering. In addition, it is hoped that the OD^3 experience and expertise in both industry and government will be made available to this program effort as it proceeds. This can be done through individual contacts, seminars, and workshops[11].

Summary

As described, storage medium lifetime is a subset of the definitions in the General life expectancy factors section of this paper. Whereas the associated systems including optical disk drives, servos, lasers, and optics are usually replaceable (providing that they do not fall into the one-of-a-kind category), data contents which are lost due to the OD^3 media failure in storage may be irretrievable. Therefore, the principal factors to be investigated by the Computer Storage Media Group are related primarily to the life expectancies of the media and media structures. The initial program efforts will be towards determining the intrinsic degradation mechanisms of the media materials through static optical, chemical, and physical testing. Subsequently, these test results will be correlated with dynamic test results derived from the system dependent degradation factors of the media. The correlation of these two test modes will serve as the background for the development of a quantitative OD^3 lifetime prediction theory, a media care and handling methodology, and finally a Standard Reference Material, if feasible. These results may also be of value in the formulation of OD^3 standards. Note, however, that media standards usually incorporate parameters which are also related to media test, operational, as well as the storage environments.

References

1.   Geller, S. B. Calibration of NBS secondary standard magnetic tape (computer amplitude reference) SRM 3200 using the reference tape amplitude measurement "process A" model II. Nat. Bur. Stand. (U.S.) Special Publication 260-29; June 1971. 49 p.

2.   Geller, S. B. The effects of magnetic fields on magnetic storage media used in computers. Nat. Bur. Stand. (U.S.) Technical Note 735; July 1972. 30 p.

3.   Geller, S. B. Erasing myths about magnetic media. Datamation; March 1976. 5 p.

4.   Geller, S. B. Care and handling of computer magnetic storage media. Nat. Bur. Stand. (U.S.) Special Publication 500-101; June 1983. 128 p.

5.   Freedman, J. B. The National Bureau of Standards/Institute for Computer Sciences and Technology (NBS/ICST) optical digital data disk (OD^3) standardization activities. International Society for Optical Engineering (SPIE). Brussels, Belgium; June 1984. 3 p.

6.   Geller, S. B. Archival data storage. Datamation; October 1974. 4 p.

7.   Goldberg, M. F.; Vaccaro, J. eds. Physics of failure in electronics. Rome Air Development Center - RADC Series in Reliability, Vol. 2; March 1964. 559 p.

8.   Pau, L. F. Failure Diagnosis and Performance Monitoring. Mercel Dekker, Inc.; 1981. 427 p.

9.   Fluck, E.; Goldanskii, V. I. eds. Modern Physics in Chemistry. Volume 2, Academic Press Inc.; 1979. 638 p.

10. Mittal, K. L. ed. Adhesion measurement of thin films, thick films, and bulk coatings. ASTM Special Technical Publication 640; November 1976. 402 p.

11. Freedman, J. B. ed. Standardization issues for optical digital data disk (OD^3) technology. Nat. Bur. Stand. (U.S.) Special Publication 500-111; April 1984. 227 p.

Session 4

Future Directions

Chairman
Masafumi Mori
Toshiba Research and Development, Japan

Mark Goldberg
Department of Defense, USA

A Digital Optical Recorder (DOR) for application in harsh environments

T.J. Stahlie and R.A. Falkenburg

DOR marketing group, Hollandse Signaalapparaten B.V.
P.O. Box 42, 7550 GD Hengelo (Ov), the Netherlands

Abstract

Digital Optical Recorders are suited for mass data logging and storage of text, pictures and other forms of non-moving images. Most of the recorders developed until now, have been designed for computer-room environments. However, a number of applications will require recorders to be used in non-benevolent environments. Hollandse Signaalapparaten B.V. (Signaal) has developed a robust industrial version of the Digital Optical Recorder and aims at the business of complete systems (including DOR as an archival medium) for storage and retrieval of data, images and documents in heavy duty and special professional applications.

Because of the system approach, emphasis is put on the development of an intelligent controller for adaptation to a variety of host computers.

Introduction

One of the latest concepts to be utilised in mass data storage technology has been recording digital information on an optical disk. Such vast quantities of information can be stored, that this innovation heralds a new era in information storage techniques. Originating from the Philips Research Laboratories, the DOR concept has been refined and developed by Signaal to produce a DOR-system for application in arduous industrial and military environments.

In designing the DOR system, it was necessary to create a careful balance between the complexity of its two main components, the standard 12 in. optical disk and the disk drive. Too simple a disk would require an expensive and complicated disk drive, while a high precision disk with a simple drive would make the cost of the disk prohibitive; an important consideration for a system using exchangeable disks. Signaal has drawn upon Philips' long experience in both Laser Vision (LV) and Compact Disc (CD), to produce a design that strikes the correct balance.

In fact, the DOR system not only opens up new areas for the application of computerised mass storage methods, but will also initiate the rethinking and development of new ideas in information storage techniques. Apart from other interesting application areas, there is a strong trend towards document storage and retrieval systems for engineering drawings even in non-computerroom and/or non-benevolent environments. Signaal developes a system philosophy for this application and pays special attention to the interface of the DOR to host computers.

The robust industrial disk drive design

The optical disk is a sealed double sided air/glass sandwich construction with the internal glass surface coated with tellurium alloy which forms the recording medium. The DOR disk is pregrooved with a spiral groove, along which the laser spots can track and stay in focus. The spiral groove is densely packed with 625 turns per radial mm. The disk is divided into sectors (Figure 1).

Each track/sector segment is preformatted with a header and a unique address. Data is written onto the disk by melting very accurately positioned holes into the metal coating along the pregroove spiral in the area after each header. Every sector can contain up to 512 bytes of user data. An in-track clock (pilot signal) is written into the bottom of the pregroove spiral, for synchronization. Both the clock track and the header/address relief pattern are pressed into the disk at the manufacturing stage.

Figure 1. The pregrooved, sectorised Philips disk.

Figure 2. The Digital Optical Recorder, Robust Industrial version.

The DOR drive (Figure 2) can be supplied either in cabinets to suit customer requirements, or in standard 19 inch rack width frame.
The disk drive consists of a drive deck and electronics rack. The drive deck contains the optical tracking and focussing mechanism, rotation motor and optical/electronic components. In the standard version the drive deck is mounted at the top of the cabinet and is protected by a dust cover which allows access to the top of the drive unit and the disk. Also "automatic" cassette exchange mechanismes are available. The electronics rack and supply systems are fitted in the lower part of the standard cabinet.

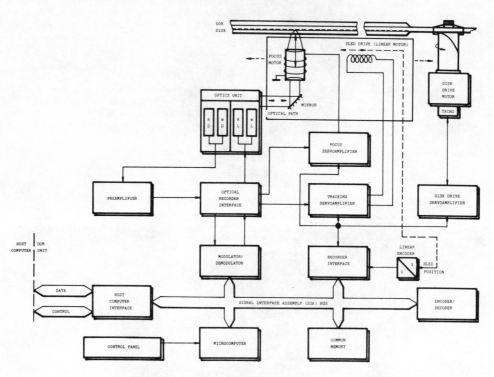

Figure 3.   DOR system block diagram.

Figure 3 shows a schematic diagram that outlines the operation of the disk drive. High speed random access for data is possible because of the use of dynamic tracking in conjunction with the preformatted spiral groove and other features of the drive, e.g. sled air bearings and linear motor used for tracking and focussing.
During a data retrieval operation the optical system uses an optical rule to obtain a course positioning so that the read light spot is brought within 40 tracks of the target track, after which the fine positioning system, which reads the sector heading, takes over to obtain the correct information address. In practice an average access time of 200 ms is obtained. High speed random access is important for any information system. Signaal's DOR-drive provides fast access in conjunction with high data reliability and very high storage capacity, which because of the inherently robust tracking and focussing systems, gives an excellent, virtually error free performance, even under demanding environmental conditions.

The DOR system laser/optics block contains two lasers, two detectors and their associated optical components. The lasers are the new CQL type heterojunction semi-conductor lasers, each of which is housed in a light pen with the necessary collimating optics. The optics block has been mounted on the baseplate, not on the moving sled. The read and write laser beams are focused onto the metal coating on the inside of the disk. Polarisation techniques and a small wavelength difference between the two laser beams ensure that the reflected beams impinge on the correct detectors. The read detector is used for data retrieval and to check data as it is written. The DRAW function (Direct Read After Write), which is used to ensure data integrity at the moment of recording, makes use of this detector set-up.
Both the read and write detectors monitor clock information that is preset in the pregroove spiral to determine precise timing.
Servo control and dynamic tracking of the sled motor can maintain an accuracy of approximately 0.1 micron.

As has been stated before, Signaal's DOR has been designed for high-industrial, non-computerroom applications with a growth-potential to specific military environments. The most important ruggedized elements in the robust industrial unit now available are:
- full mil-spec Signaal multilayer PCBs. The packing density of a Signaal multilayer PCB equals approximately the double Eurocard capability. However, the Signaal PCB is smaller and is highly reliable even under adverse environmental conditions.
- application of air-bearings for sled-movement and lens-movement. This results in a vibration and shock resistant tracking system.
- shock-absorbing mounts at the top of the cabinet for shock-protection of the drive deck.

A number of experimental tests have been carried out to investigate the ruggedized characteristics. The most important one is the DEF-133 shock- and vibration test, which led to the conclusion that the unit has unique characteristics in this field making it ideal for shipborne and land-mobile applications.

Finally, Table 1 gives an overview of some of the most important system specifications.

Table 1. DOR-RI system specification

disk	Philips air-sandwich glass disk
storage capacity	1000 Megabytes of user data per disk side
sector capacity	512 user bytes
disk diameter	12 in. nominal
track separation	1.6 μ
disk drive	Signaal; Robust Industrial version
solid state laser	CQL-type
read laser	continuous output 2 mW
	wavelength 870 nm
write laser	standby cont. output 1 mW
	pulsed uptr 20 mW
	wavelength 820 nm
sector error rate	10^{-12} or better
access time (average)	200 ms

Host computer interface

In order to make DOR units more user-friendly, Signaal has decided to develop an intelligent interface to facilitate the connection to a variety of host computers. This interface is realized in firmware and consists basically of four extra PCBs to be added to the electronics rack of the standard DOR.
This interface is called Recorder Interface Processing Unit (RIPU); added to the standard DOR the unit is called DOR-RIPU.

In the DOR-RIPU configuration the command- and databus are separated. This opens the possibility to bypass (and thus unload) the host computer for data transport. The data traffic is routed over a 16 bit parallel unidirectional asynchronous interface. The command path goes over a serial communications link according to the V24 or RS232 hardware interface.

RIPU performs the following functions:
- acting as the interface of the standard DOR to the host;
- buffering data to meet the demand of timing of the DOR during reading and writing;
- performing all device management;
- performing all volume management;
- performing all file management.

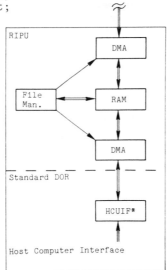

Figure 4. DOR-RIPU intelligent interface.

The four extra PCBs are the following ones (Figure 4):
A DMA card contains the interface to the standard DOR (based on DEC DMA). The RAM card contains 256 Kbyte of dynamic RAM, sufficient to contain 512 data blocks of 512 bytes each and acts like a data buffer. The micro processor card contains a 6809 processor type of Motorola together with 12 Kbyte of ROM and some 2 Kbyte of RAM. A management service micro-software package will be loaded as firmware on the ROM. The RAM is filled with information on the device, the mounted volume and the files on the volume. When a user opens a file for read and write actions, RIPU copies the file information from the disk into the RAM. A maximum of 8 files may be opened at one time. For the remaining host DMA card several versions are available, each suited for a type of host computer to which the DOR has to be connected.

Apart from this RIPU firmware solution for the interface, Signaal can also supply a driver and file management software package. However, such a package is only available for VAX computers with the VMS operating system. The software package, called DORIOS, performs the same functions as RIPU does. In this case, the standard DOR unit can be selected and because the interface of the standard drive is based on DEC-DMA, a connection to a VAX machine is very easy. With DORIOS, the user is able to use at application level simple commands for opening files, closing files, reading and writing.

DOR applications

The total newness and uniqueness of DOR not only opens up a vast array of new applications for mass data storage, but also expands the possibilities for existing systems. DOR's very high data density and robust design combined with the tamper-p oof nature and long working life of its recording medium have resulted in a system with many advantages over existing data storage systems. It has an enormous storage capacity, but with manageably sized equipment and a competitive price per bit stored, while maintaining a high degree of data security even when used in harsh environments.

The introduction of DOR has permitted computerised data storage techniques to be used for applications and taken into situations where it was not previously possible. Ideas in this field are already changing so that existing and future mass data storage systems, using DOR, can realise their full potential. Therefore, as an aid to the visualisation of possible applications for the DOR system, two broad categories have been defined:
. mass data logging
. document storage and retrieval.

Mass data logging

There are many applications which generate vast quantities of data that have to be recorded and stored for analysis at a later date.
Reliable and accurate recording can be extremely important especially where the project or event cannot be repeated due to the impossibility or the expense of repetition. Geological and seismic surveys produce large amounts of data which have to be recorded in various types of difficult environments that can range from arctic to desert conditions. The explosion used during seismic surveys lasts only about 6 seconds, but from the shock waves produced large amounts of data are obtained about the surrounding rock structure.
Data from geological and seismic surveys must often be kept for long periods of time. DOR, with its large data storage capacity and very long storage life is ideal for this type of application, providing a significant improvement over other methods.

Complex measurement systems such as those used in equipment trials are examples of mass data logging. Over short periods of time large amounts of data have to be recorded accurately. Quite often these tests can only be completed once so that the data must be recorded correctly the first time. If these trials are on ships or oil rigs etc., testing may be underwater or in harsh environmental conditions. Signaal's DOR with its robust recorder and environment proof disk has obvious advantages over other systems under these conditions.

Other mass data logging applications include those where large amounts of data must be collected and stored but where the data is continuous and is used to monitor certain types of operations. Metering for oil-flow, in-service inspection of nuclear power stations and satellite data dumps are examples of these applications.

It is clear that DOR can only be integrated in these measurement systems if the interface problem is solved adequately. That is the reason that Signaal offers the solution of the interface together with the delivery of the disk drive.
DOR-RIPU as described before is an ideal concept for this purpose.
As such we have delivered even a drive system with an interface to a special military computer system. This illustrates that even special interface problems can be solved.

Document storage and retrieval

This category includes applications where large amounts of information, available in the form of magazines, manuals, pictures, drawings, etc., need to be stored and when required, must be easily retrievable. This information is, in general, not available in digital form, and therefore, it must be digitised using scanners. Even after data compression techniques are applied, the large volume of data obtained from this digital conversion means that extremely high capacity data storage with rapid random access is necessary. A single sided DOR disk can hold up to about 30.000 pictorial, black and white A4 pages. If the information consists entirely of alphanumeric characteristics, for example from a word processor, then up to 500.000 pages can be stored.

A typical area where DOR is extremely beneficial is the presentation of (full-colour) maps on screens. Charts and maps can be digitised and each full sized, multi-coloured chart or map can require, in un-compressed format, 20 Megabytes of data storage. Experiments within Signaal has shown that sea-charts for instance can be "reproduced" on a high-resolution screen with all the details of the original document. A map with the size of the screen will take 2.5 Mbyte storage space, resulting in the storage of 800 of those maps on a DOR disk.
This application is very interesting for military users, but also for users of maps in other fields.

DOR also has distinct advantages for the storage of servicing documentation. Manuals and engineering drawings up to A0 size can be stored on the DOR system with a significant reduction in space required for this material compared to paper based or aperture card type systems. A further advantage is that the order of the stored material and the material itself is not disturbed during any data retrieval operation.
These features combined with DOR's robust nature make it ideal for mobile, shipborne, or off-shore applications where space may be limited and environmental conditions can be harsh.

Storage and retrieval of engineering drawings is becoming more and more a problem for users on board of ships, off-shore platforms, but also for users in high industrial production companies. Signaal has developed a system concept for storage and retrieval of this type of documentation. The DOR is within this concept the key element for data storage.
Input scanning is recommended via an aperture card scanner which can scan at a speed of three cards per minute. The scanned information is written on the disk, while the file administration takes care for the address labels. The stored data can be used for presentation on high resolution terminals where the retrieved process is carried out by the operator. After declaration of the selected output portion the information can be produced by a hardcopy unit or by other output plotters. Figure 5 gives a generalized picture of the system concept.

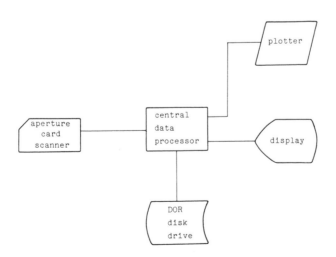

Figure 5. Generalized system concept for storage of engineering drawings.

Other examples in document storage and retrieval applications are storage of X-rays or other images which may only be used once, but must be kept for longer periods of time, due to - f r instance - legal requirements and yet be easily accessible if required. DOR is also ideal for this type of storage enabling data to be stored for very long periods due to the stability of the recording medium. Also with the DOR system infrequently used information will not become unaccessible and difficult to find.

Since Signaal is a supplier of very complex systems such as Air Traffic Control systems or Command and Control systems and accepts overall system responsibility, delivery of technical documentation is a very critical and important aspect for the company. Trends are going into the direction of electronic storage and distribution of this type of information. Even on board of ships it will be very practical in the future to retrieve the relevant documentation for operation, service or repair. System concepts are also under development in this area.

Concluding remarks

A ruggedized Digital Optical Recorder is currently available from Signaal. The system meets specifications for non-computerroom applications. Signaal has a tradition to make customised systems to suit any particular application. As such, the rugged characteristics can be extended to special drive characteristics even suitable for military applications. Furthermore, Signaal is prepared to take responsibility for building systems around the DOR drive, or to integrate DOR into existing systems. Systems produced can be tailored to customer requirements and can include additional items, such as high-resolution displays, high- or medium-resolution scanners, data transmission facilities and application software. In particular attention is paid to solving the interface DOR-host computer. Intelligent controller facilities for drives and file management functions are available.

Optical disk recorders for operationally demanding mass storage applications

Gerald M. Claffie

RCA Government Systems Division/Advanced Technology Laboratories
Building 10-8-1, Delaware & Cooper Streets, Camden, New Jersey   08102

Abstract

The design approaches underway at RCA, as well as the performance parameters being addressed, are described for both argon-laser- and semiconductor-laser-based recorders. These recorders have been under development at RCA since the early 1970's for operationally demanding user requirements that cannot be suitably addressed by any commercially available optical storage system. Current and projected characteristics of the disk media used with these recorders are also described.

Introduction/background

In the optical disk recording market, RCA addresses the specialized user requirements segment that cannot be satisfied on either technical merit or cost effectiveness by recorders intended for the more general-purpose commercial market. Both argon-laser- and diode-laser-based recorders, laser diodes, monolithic laser-diode arrays and monolithic detector arrays have been developed as part of this effort. Until recently, this recording hardware and device development work depended almost completely on research samples of optical disk media provided by the RCA Laboratories[1]. Current efforts are directed toward hardware design approaches that are compatible with cost effective, commercial media.

The argon-laser-based optical disk work has concentrated on addressing very high user data rate and disk storage capacity requirements and automated, rapid access to the data on disks stored within an on-line storage library, or "Jukebox". These requirements apply today primarily within current and planned non-mobile government data-processing and storage facilities. In response to these requirements, the Air Force (Rome Air Development Center) has sponsored development programs at RCA to demonstrate (1) single-channel recording at user data rates to 50 Mbps, (2) multiple-channel recording at user data rates to 300 Mbps, and (3) data packing densities on the disk surface compatible with achieving 10^{11} user bits per side of a 14-inch disk. Most recently, under a program sponsored by the Air Force and NASA, a "jukebox" mass memory system is being developed to provide, in less than 6 seconds, access to 10^{13} on-line user data bits. Two engineering models of this system will be delivered to testbed facilities in 1984; one to the Rome Air Development Center and the second to the NASA Marshall Space Flight Center. Experience data from these testbed systems will be used to determine what design refinements are required for the system to meet user reliability, maintainability, and availability requirements. Current efforts are also directed toward the rapid exchange of disks between the "Jukebox" and a remote storage library that can provide automated access to a virtually unlimited supply of on-line data.

Diode-laser optical disk recording work is based upon research and process development work at the RCA Laboratories. Recent work at the laboratories has been directed toward increasing the power output and the reliability of single-mode lasers and on the development of monolithic laser-diode arrays. For example, constricted double-heterojunction large optical cavity (CDH-LOC) lasers have been developed to provide single-spatial mode output with a peak power of 40 mW in pulsed operation at a 50% duty cycle. These devices have been operated for over 10,000 hours with small drops in output power and minimal change in beam shape.

Recorders have been under development since the late 1970's to use single laser-diode sources, with recent work also addressing the use of monolithic diode arrays of up to 10 diodes. The objectives of current programs are to increase the data rate and disk data packing density that can be achieved with single-diode and diode-laser array recorders and to address the packaging, reliability, and maintainability issues that apply to compact, ruggedized recorders intended for use in a non-benevolent operating environment.

Argon-laser-based recorders

The key user requirements that have been addressed by argon-laser-based recorders are shown in Table 1. These recorders are typically configured as a Disk Drive Unit containing the electronics, optics, and mechanisms required to write to, read from, and handle the optical disk media and a Controller that controls and monitors the Disk Drive Unit operations and interfaces the recorder with the host system. A block diagram of the Disk Drive Unit for a single turntable recorder with a single write/read optical head and automated disk

Table 1. Key User Requirements-Argon Laser Based Recorders

Requirement	Current/Past	Projected
Continuous Data Rate (Mbps)	0-300	0->300
Beams/Optical Head	9	9
Disk Capacity (bits)	$\sim10^{11}$	$\sim5 \times 10^{11}$
14-inch diameter	-single sided	-double sided
	-CAV Recording	-CLV Recording
Read Bit Error Rate	$<10^{-9}$	$<10^{-12}$
Data Access Time (seconds)	<0.5	<0.25
For mounted disk		
On-Line Disk Access (seconds)		
-Worst Case (100 disks)	<6	<6
-Disk in queue position	<3.5	<3.5
-From Storage Library	N/A	<15
Operating Environment	Fixed-Benevolent	Same

handling capability is shown in Figure 1. A block diagram of the associated Controller is shown in Figure 2.

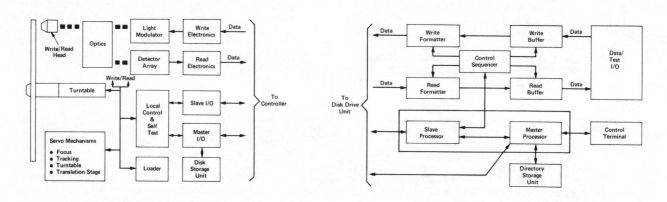

Figure 1. Disk Drive Unit

Figure 2. Controller

Controller functions

The Controller is intended to communicate, via a control command and status link, with the host system and to input/output high-rate user data over separate channels. The Controller design is modular so that it can be expanded to handle multiple data channels per write/read head, multiple heads per turntable and multiple turntable Disk Drive Units. Flexibility and adaptability to varied user requirements is also achieved by using ROM-based microcode-driven write/read data processing hardware and ROM-based software-driven master and slave control processors[2].

The write and read buffers provide a storage capacity equal to two disk revolutions, allowing asynchronous data transfer between the host system and the Disk Drive Unit. A continuous user data rate of up to 33.3 Mbps per channel can be accommodated with a burst rate up to 80 Mbps per channel until the buffer is full. Each user data channel is 8 bits wide to maintain a user word rate below 10 Mwps. The write formatter adds overhead information for user data error protection and identification. Three levels of error protection are provided: (1) write verification with rewrite, (2) short burst EDAC encoding for random noise and small defects, and (3) long burst EDAC encoding for scratches and other large dropouts encountered during storage and handling. The write formatter output is transferred to the Disk Drive Unit as separate 8-bit wide channels corresponding to each user channel. The read formatter is used in the write verification process and accepts read data from the Disk Drive Unit. The read data errors are detected and corrected, the read data is identified, and the read data is output to the read buffer.

The master and slave control processors perform all functions needed to control the optical disk recorder. Time critical control of the write/read functions requires that one slave processor be dedicated to each turntable position. Multiple turntable Disk Drive Units require multiple slave processors, all communicating with a single master. The master processor functions as the user command interpreter, the disk location directory, the mounted disk data directory, and controls operation of the disk storage unit.

Disk Drive Unit functions

Write electronics. Error protected user data is input to the multi-channel write electronics, with each channel capable of accommodating a data rate up to 66.6 Mbps during the active write/read operation. The 8-bit wide data for each channel is combined into a serial data stream, encoded into a run length limited code to take advantage of the characteristics of the optical write/read process[3,4], and applied via high-speed emitter-coupled logic to a multi-channel acousto-optic light modulator. Up to nine write channels have been implemented to accommodate a continuous user data rate of 300 Mbps with a single write/read optical head.

Write/read optical system. A schematic of a two-beam write/read optical system is shown in Figure 3. Systems of this type have been built for up to nine beams per optical head[5]. Major subassemblies include the argon laser, the stationary optics, and the moving optics on the write/read head translation stage. Briefly, light from the argon laser passes through a half wave plate and reflects off two mirrors that adjust beam position. From the second mirror, the light is passed through two beam splitters, where a portion of the beam is split off for the read and pointing monitor functions and to the write and read phase gratings. The write phase grating splits the light into two divergent beams, which the following lens then images into spots under the modulator transducers. The diffracted light from the modulator is collimated by the next lens and passes through the Dove prism, which is used to vary the spacing of the tracks formed on the disk by rotating the image of the two write spots. The collimator expands the beams from the modulator to fill the entry pupil of the objective lens and relays an image of the grating (the "source" of the two spots) into the objective lens to form the write spots on the disk. The format of the data on the disk surface achieved with this optical system for the NASA Recorder is shown in Figure 4. For this application, 16 tracks are written (8 groups of 2 tracks per group) for each successive 20-µm write/read head translation stage position by step positioning the galvanometer mounted near the objective lens.

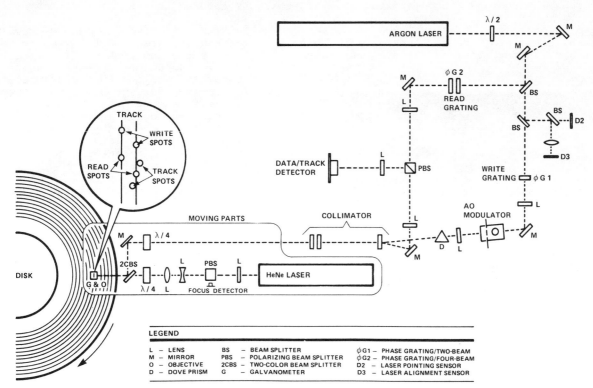

Figure 3. Schematic of typical multichannel optics (argon laser)

The read beam at the input to the two phase gratings is split into six beams (an array of 3x2 beams), four of which are used. These beams are imaged near the polarizing beam splitter and recollimated by the next lens to pass through the collimator and the objective lens to the disk. Light reflected from the disk passes back through the optics to the polarizing beam splitter. The beams are reflected to the detector since the double pass through the quarter wave plate has rotated the plane of polarization 90 degrees. The 9-avalanche detector, 4-PIN detector monolithic array developed by RCA for use in multiple beam optical disk recorders is shown in Figure 5. The high-gain avalanche detectors are

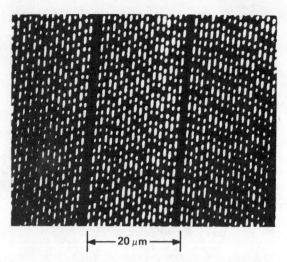

Figure 4.   Track groups on disk

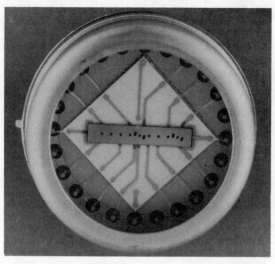

Figure 5.   Monolithic detector array

for read signal detection; the PIN detector outputs are applied to the data tracking control loop.

Light from the HeNe objective lens focus laser is focused, then expanded by two lenses, and passed through the two-color beam splitter to the galvanometer mirror and the objective lens. Light reflected from the disk passes down the optics chain and, by the action of the quarter-wave plate and the polarizing beam splitter, is coupled to the focus detector.

Read electronics. The equalized detected read signal for a 57-Mbps channel is shown in Figure 6. This photograph illustrates the high signal-to-noise ratio and the clearly defined "eye" pattern that is achieved at the very high data rates and disk packing densities of interest. The signal shown in Figure 6 is hard limited and applied to a clock recovery phase-locked-loop for strobing to recover the digital data stream, to the synchronization pattern detection logic, and to the data decoding logic. The decoded data for each read channel is demultiplexed to an 8-bit wide data stream and output to the Controller.

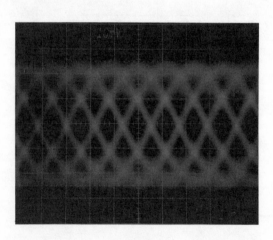

Figure 6.   Detected read signal

Servomechanisms. The very high disk user data capacity cited in Table 1 is obtained by efficiently encoding the write data and by achieving a minimum in-track recorded feature size of 0.42 µm and a track-track separation of 1.0 µm. High performance focus, tracking, and translation-stage position control servomechanisms are required to meet the read bit error requirement at this high disk data packing density. For example, the write/read objective lens is moved as the disk rotates to maintain focus to within 0.2 µm, the read beams must follow the recorded data tracks to within 0.1 µm, and the translation stage must maintain position stability to within 0.1 µm. The turntable servomechanism maintains a single rotation rate for constant angular velocity (CAV), recording with this rate determined by the user data rate per channel, the minimum recorded feature size, and the smallest disk recording radius to be used. A more complex turntable servomechanism is required if constant linear velocity (CLV) recording is used to increase the disk capacity by reducing the disk rotation rate as the recording radius is increased.

Recording and disk handling mechanisms. The optical write/read head translation stage, the "Jukebox" configured disk handling mechanisms, shown in Figure 7, and the protective disk cartridge shown in Figure 8, have a strong influence on the User Requirements listed in Table 1.[6]

The translation stage on the Air Force and NASA recorders is required to jump 20 µm to within 0.1 µm in less than one disk rotation so that the continuous user data rate per channel can be maintained when using a concentric disk data format at a 50 percent

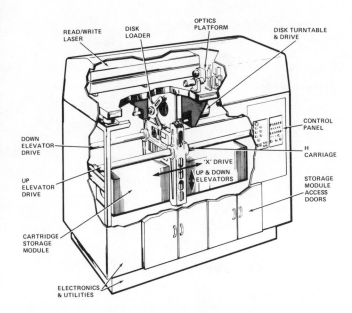

Figure 7. Conceptual drawing of
Disk Drive Unit

Figure 8. Disk in protective
cartridge

write/read duty cycle. The concentric format facilitates short write sessions where user data is asynchrously input to the Controller write buffer until the data for a full disk rotation is accumulated. The concentric format also facilitates rapid access to stored data; a worst-case translation stage jump time of 0.4 s provides worst-case data access within 0.5 s. Alternately, a spiral recorded data format may be implemented for data capture applications where the peak disk write/read rate is continuously required for multiple disk revolutions and less rapid read access to the data is required.

The storage module shown in Figure 7 will accommodate 128 on-line, 14-inch disks in protective cartridges. The cartridges are elevated from storage to the translation position, moved laterally to the queue position, elevated to the load position, and mounted on the turntable. A top view of the turntable/translation stage assembly with a mounted disk in its protective cartridge is shown in Figure 9.

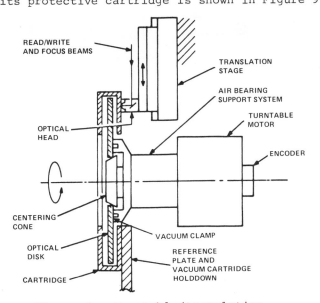

Figure 9. Turntable/translation
stage assembly

The cartridge facilitates manual and automated disk handling and storage. The disk is suspended and locked in the cartridge at all times except when mounted on the turntable to prevent contact between the cartridge and the write/read surface. The 6-s worst-case access time to the data from an on-line disk applies at system turn-on or periodically when the sequence of write/read operations is not continuous. A dual elevator mechanism is used so that the next disk of interest can be queued beneath the turntable to shorten the inter-disk data access time to under 3.5 s.

Diode-laser-based recorders

The key user requirements that are being addressed by diode-laser-based recorders are shown in Table 2. To date, work has concentrated on developing the 2-diode Disk Drive shown in Figure 10 that can accommodate a monolithic array of up to 10 diodes. The Disk Drive design approach plans for installation in a ruggedized package (of the type shown in Figure 11) that includes all optics, mechanisms, and electronics required for a manually loaded single-disk recorder. The ruggedized recorder requirements include a package size under 12x16.5x28 inches and a weight under 150 pounds. Continuous user data at rates up to 20 Mbps is accommodated using a concentric disk format with a 50 percent write/read duty cycle. Each CDH-LOC diode in a 2-diode array operates above 20 Mbps during write/read to accommodate the user data plus overhead. The two-channel operation allows closer track positioning than is possible

Table 2. Key User Requirements—Diode-Laser-Based Recorders

Requirement	Current	Projected
Continuous Data Rate (Mbps)	0-20	0->100
Beams/Optical Head	2	~10
Disk Capacity (bits)	$(2-5) \times 10^{10}$	$(1-2) \times 10^{11}$
14-inch diameter	-single sided	-double sided
	-CAV Recording	-CLV Recording
Read Bit Error Rate	$<10^{-8}$	$<10^{-12}$
Data Access Time (seconds)	<0.5	<0.25
For Mounted Disk		
On-Line Disk Access (seconds)		
-Worst Case (100 disks)	<6	<6
-Disk in Queue Position	<3.5	<3.5
-From Storage Library	N/A	<15
Operating Environment	Ground/Air	Aircraft
	Transportable	Spacecraft
		Operational

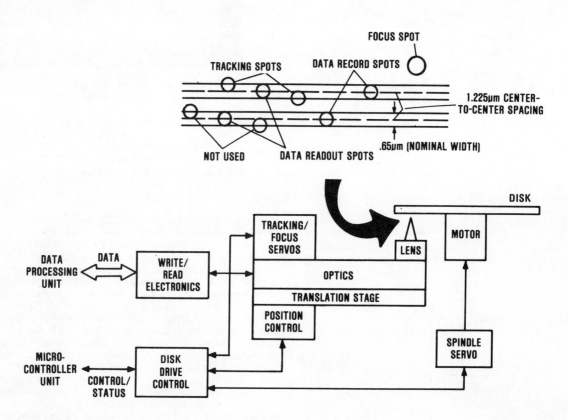

Figure 10. Disk Drive Unit (2 laser diodes)

with single diode sources by reducing the number of guard-bands that are required when the write beams are positioned between disk rotations. This is an aid in achieving the disk capacity goal of 5×10^{10} user bits per side of a 14-inch disk. A HeNe laser beam is split by cascaded transmission phase grating into multiple beams for dual-track readout, focus control during write/read, and radial track control in read. The optical footprint at the disk plane is shown in Figure 10.

The 2-diode array has been produced by the RCA Laboratories as a step in the develop-ment of a monolithic array of 10 or more high-power, single-mode diode lasers which can be modulated directly and independently[8]. The diodes within this array, which are located on 150-μm centers, are intended to output more than 20 mW at 50 percent duty cycle at a re-liability comparable to single-diode devices. Preliminary recording results using three adjacent diodes within a 10-diode array are shown in Figure 12. The nominal track spacing is 1.75 μm. The upper and lower tracks are 10-MHz and 12.5-MHz square waves, respectively. The middle track is FM-encoded video with an 8-MHz carrier frequency.

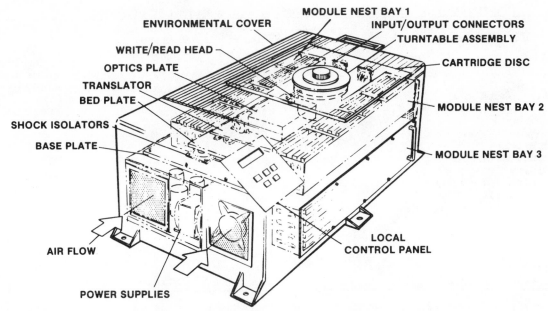

Figure 11. Conceptual drawing of ruggedized recorder

Recording media

The current and projected media requirements for the recorders being developed at RCA are listed in Table 3. Most of the work to date has been based upon the use of unformatted, single-sided, archival media on 14-inch aluminum (Winchester) magnetic-disk substrates. The recording media research undertaken by the RCA Laboratories to support the recording hardware development work has been vital — samples from this work were at times the only media available to evaluate recorder performance. However, principal long-term goals have been to be as compatible as possible with evolving media, to aid potential media suppliers by documenting requirements in an interactively defined specification[9], and to test candidate production media samples as available.

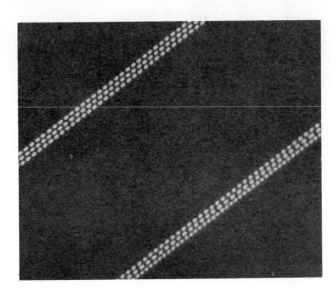

Figure 12. Three-channel recording with monolithic diode-laser array

Media issues of current interest include the approaches to disk mounting eccentricity correction for both formatted and unformatted disks, the effects of using a flexible disk substrate, the availability of double-sided disks, and the availability of reusable (erasable) media. All of this media-related work is intended to assure that the recording hardware will meet operationally demanding user performance requirements with the cost-effective use of production media.

Potential for growth in capability

Very high user data rates above 300 Mbps per write/read head can be accommodated by the multiple beam, argon-laser-based optical system (9 beams/head); this system allows multiple write/read and read-only heads to be serviced by a single laser. Also, user data rates to greater than 100 Mbps are anticipated for a single write/read head using a monolithic array of 10 diodes.

The automated disk handling capability that has been developed is compatible with an argon laser, a diode laser, or a diode-laser array as determined by user data rate, disk capacity, and operational environment requirements. The modular electronic, electro-optic, and mechanism design approach taken for these recorders provides for a convenient expansion in capability to multiple turntables serviced by a single on-line disk storage rack and to multiple write/read or read-only optical heads with each turntable. The horizontal axis of

Requirement	Current (Argon/Diode Laser)	Projected
Table 3. Summary of recording media requirements

Requirement	Current (Argon/Diode Laser)	Projected
Disk Diameter (inches)	14	5.25-14
Substrate	Winchester (Aluminum)	-Aluminum -Flexible
Rotation Rate (rpm)	<3600	<3600
Operating Wavelength (nm)		
-Write	488/820	488/<820
-Read	488/633	488/633/<820
-Focus	633	633
Write Spot Size (μm)	0.4/0.65	0.4/0.65
Write/Read Data Rate (Mbps)	<67	<100
Disk Sensitivity (mW)	5-15	5-15
-Surface Lineal Velocity (900-1400 in/s)		
Read Carrier-to-Noise Ratio (dB) -22 MHz Tone/30 KHz BW	50-60	50-60
Read Bit Error Rate (BER)	$<10^{-4}$	$<10^{-5}$
Archival Media Lifetime (years)		
-Prior to Use	4	4
-After Use	TBD	10-20
Reusable (Erasable) Media	Evolving	Yes

disk rotation, with vertical disk write/read and storage, provides essentially identical access to the second side of a double-sided disk and facilitates the automated exchange of disks between the on-line storage rack and an external storage library. This provides for the on-line storage of 10^{15} or more user data bits in a facility of the type conceptually shown in Figure 13.

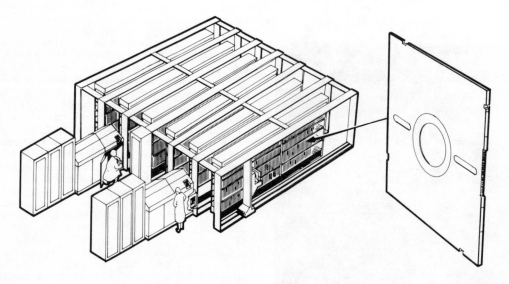

Figure 13. External storage library

Conclusions

The modular, adaptable design approaches that have been taken to develop argon laser- and diode-laser-based optical disk recorders have been described. These recorders are intended for operationally demanding user applications that cannot be technically or cost effectively addressed by recorders intended for general, commercial use. Compatibility with the use of production archival or reusable media is planned to reduce user operational costs.

References

1. R. A. Bartolini, "Media for High Density Optical Recording," Optical Recording, Vol. 20 No. 3, May/June 1981.
2. G. J. Ammon and J. H. Hoover, "A 10^{13} Bit Optical Disk Recording System," Proceedings Sixth IEEE Symposium on Mass Storage Systems, Vail, Colorado, June 4-7, 1984.
3. R. J. Tarzaiski, "Selection of 3 Phase (1,7) Code for Improving Packing Density of Optical Disk Recorders," Proceedings Optical Disk Systems and Applications, SPIE Vol. 421, June 1983.

4. P. C. Hess and J. L. Waring, "Implementation of the 3 Phase (1,7) Code for Improved Packing Density on Optical Disk Recorders," SPIE Vol. 421, June 1983.

5. C. W. Reno, "Optical Disk Recording Techniques for Data Rates Beyond 100 Mbps," SPIE Vol. 421, June 1983.

6. G. J. Ammon and B. W. Siryj, "Automated Handling Mechanisms for an Optical Disk Mass Memory System," SPIE Vol. 390-3, January 1983.

7. A. M. Earman, "A High Performance Optical Disk Drive Unit for Use in Non-Benevolent Environments," SPIE Vol. 421, June 1983.

8. D. B. Carlin et. al., "Multichannel Diode Laser Array Optical Recording," <u>Proceedings OSA Topical Meeting on Optical Data Storage</u>, Monterey, California, April, 1984.

9. M. L. Levene, "Optical Disk Media Parameters and their Relationship to Equipment Design," SPIE Vol. 421, June 1983.

Self-Coupled Optical Pickup (Ⅱ) and Its Application

Yoshinobu Mitsuhashi and Junichi Shimada
Electrotechnical Laboratory, Tsukuba Science City, 305 Japan
Shigeru Kogure
Suwa Seikosha Co., Ltd., Suwa-shi,392 Japan
Yorihiko Sakai
Dai Nippon Printing Co., Ltd., Shinjuku-ku, Tokyo,162 Japan

Abstract

Our paper consist of two parts. First, we have demonstrated that the characteristics of a self-coupled optical pickup are remarkably improved by use of a newly-developed so-called incoherent laser diode made by Sharp. Servo signals for focusing are obtained by wobbling method with signal to noise ratio more than 30 dB. Second, we describe a simple real time method for sensitivity measurement of optical storage media utilizing a self-coupling effects of a GaAlAs laser diode.   We have shown that we make clear the dependences of duration time in writing laser pulses and substrate materials on the sensitivity of a single tellurium film recording material.

Introduction

It is well known that with an optical feedback a laser diode (LD) exhibits the self-coupling effects, which appear as a increase of output power, a change of the voltage across the LD, etc.[1] It has proposed and demonstrated that these effects are exploited in a scheme named SCOOP to read out the optical memory. In our previous paper[2] we have shown that the operational instabilities caused by the coherent feedback must be solved in order to realize a practical application of a SCOOP. In the first part of this paper, we report that the characteristics of a SCOOP for an optical disk pickup are greatly improved using a so-called incoherent LD which is a short cavity VSIS (V-shaped substrate inner stripe) laser diode with self oscillation made by Sharp. In our experimental setup the distance between an LD and an optical disk was as short as 2.8 cm. A compact size of the pickup and its simple structure is the most significant feature of a SCOOP. There were found no problems related to the coherent optical feedback.

It is important to measure the recording sensitivities for developing new optical storage media, for example, with low cost or with nontoxicity and for evaluating a given media. Observation of the hole formation process in recording is another interest for analyzing the characteristics of the media.[3]   In the second part of our paper we show that a self-coupling effect of an LD provides quite useful technique to measure the sensitivity of optical storage media. The output light of a self coupled LD changes according to the intensity of the fed back light which is proportional to the media reflectivity. Thus, in the writing process using pulse output light of an LD, the transient changes of the reflectivity ,of an ablative optical storage medium can be observed in real time fasion.   We show that we successfully make clear the dependences of the duration time of a writing laser pulse and the influences of plastic substrate materials on the sensitivity for an optical recording material of a single tellurium film.

Self-Coupled Optical Pickup

Fig. 1 shows the experimental setup of a SCOOP. A package of an LD (Sharp Model LT-023MC) was composed into a holder together with a lens.   The output light of the LD was focused into about 2 μm diameter on the back surface of a piece of a Compact Disc substrate which was mounted on a voice coil actuator.   The numerical aperture (NA) of the lens is 0.23 for the LD side and 0.45 for the disc side respectively. The LD was operated either by a constant current (CC) source or by an automatic power control (APC) driving unit. The continuous wave oscillating threshold current was about 55 mA at the room temperature. The prominent feature of the LD is its spectra as shown in Fig. 2, which was measured by a spectrometer (Jobin Yvon Model, H. R. 1000) with a resolution of 0.08 Å.   One can note that the LD oscillates in a multi longitudinal modes, each of which has a broad linewidth having more than 1 Å.   These features are reported to be originated in self oscillation resulted partly from a short cavity structure.[4] Fig. 3 shows the characteristic curves between the signal to noise ratio (S/N) in the output light and

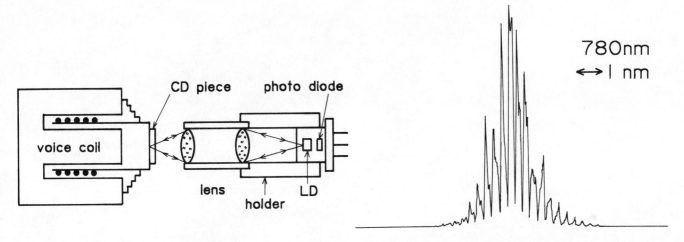

Fig. 1 Experimental setup of a SCOOP.

780nm
↔ I nm

Fig. 2 Spectra of an LD with self oscillation.

the injection current with and without an optical feedback. The noise was measured by a spectrum analyser (Sony Tektronix Model 7L13, plug-in-unit) varying the center frequency with a fixed frequency-width of 30 kHz. The curve marked with an optical feedback shows the curve in the worst case when the center frequenccy for noise measurement was varied from 100 kHz to 5 MHz, the environmental temperature was varied from 16 ℃ to 50 ℃ and the optical feedback ratio was varied from 0.001 % to 30 %. One can note that with an optical feedback the S/N is decreased as usual but the amount of S/N decreasement is only 4 to 8 dB at the injection current of 1.1 times the threshold which corresponds to the output power of 1 mW and when the injection current becoms more than 1.16 times the threshold (2 mW output power and more) the S/N is kept always more than 90 dB. In a conventional LD we used to be encountered with a coherent optical feedback which usually cause the S/N decreasement as much as 30 to 40 dB. Therefore a VSIS LD is said to be quite useful for a SCOOP scheme.

Fig. 4 shows the output light changes caused by an optical feedback with a parameter of the distance between the optical pickup and the disc at around just-focus condition. Fig. 5 shows the maximum amount of the output light changes (ΔL)max with a parameter of the injection current. In a SCOOP scheme we utilize a wobbling method to obtain the servo signals both for tracking and focusing. Usually the wobbling frequency is a few ten kHz and the noise level around those frequency is shown in Fig. 6. From these results we may estimate the signal margin for servo signals ; for an example, if we wobble a SCOOP with a ± 0.5 μm displacement at the position of a 0.5 μm out of focus with a wobbling frequency of 35 kHZ, we obtain the signal level of about - 62 dBm while the noise level at that frequency is - 90 dBm, thus the focusing signal is about 30 dB higher than the noise and will be enough as a servo signal. As for a tracking servo signal we can obtain it by superposed wobbling method with a different frequency from that for focusing.

Meantime, a radio frequency signal, which corresponds to an existence or an absence of a pit, in other words the information stored in an optical disk,is estimated about - 42 dBm in the case that (ΔL)max is 5 mV. The noise level in the frequency range from a few hundred kHz to a few MHz is less than - 100 dBm at the injection current more than 1.1 times the threshold. Therefor the above estimated radio frequency signal will be enough to reconstruct the pit information.

Fig. 7 shows the waveforms of the ouput light of a SCOOP for a ±1 μm displacement and 100 Hz wobbling frequency with a parameter of the focusing position. The signals were detected by a monitor photodiode placed within the laser package at the rear side of an LD chip. We note that at a just focus position (Fig. 6 b) the output signal is small and its waveform oscillates two times of the wobbling frequency and at an out and in focus position (Fig. 6 a and c) the signals are large and the phase of each waveform is inversed. These results are quite reasonable and imply a practical performance of a SCOOP.

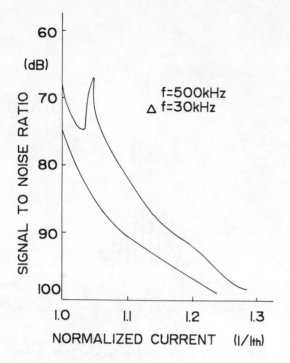

Fig. 3 Signal to noise ratio versus current.

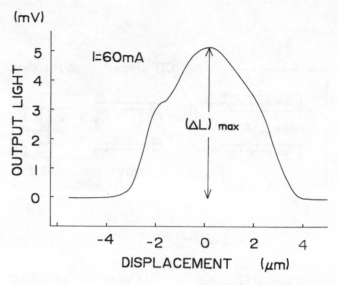

Fig. 4 Output light changes versus distance.

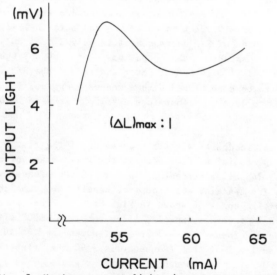

Fig. 5 Maximum output light changes versus current.

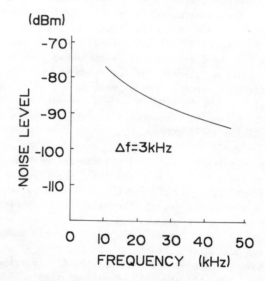

Fig. 6 Signal to noise ratio at wobbling frequency.

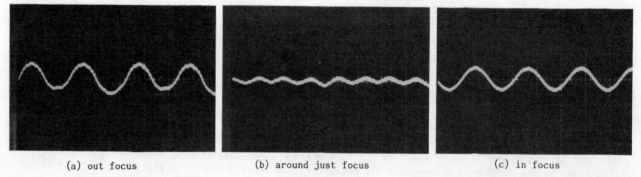

(a) out focus   (b) around just focus   (c) in focus

Fig. 7 Waveforms of output light in wobbling with a parameter of focusing position.

In conclusion we have shown and demonstrated that if we use a so-called incoherent LD with broad line-width-multimode oscillation, we could obtain sufficient wobbling servo signals and radio frequency signals in a simple and compact SCOOP scheme.   The weight of the lens and the packaged LD used in our experiment was 1.6 g and 1.3 g respectively.   The weight of a lens can be considerably decreased by a new design principle having a point-to-point imaging characteristics which is suitable for a SCOOP. The weight of an LD will be decreased by removing a window glass and the sealing package. Then, the total weight of a SCOOP will be decreased as much.   A light weight optical pickup will be quite useful for realizing high speed access and for developing a plural optical pickup system.

<h3 style="text-align:center">Sensitivity Measurement</h3>

In optical recording technology large amounts of information are stored either on an optical disc or an optical card. In these optical recording media the actual bits are mainly micromachiened with a focused laser in an optical sensitive material.   There are reports of dynamic testing of optical strage media pit formation.[5] Usually an LD is used to make micron size holes with pulsed operation and while the development of these holes is monitored with a coincident cw He-Ne laser beam.

In this paper we describe a simple method to measure the sensitivity of optical storage media using a self coupling effects of an LD which is able to observe pit developing process in real time.

The experimental setup is similar with a SCOOP as shown in Fig. 8.   If we use a SCOOP in recording process with high power output operation of an LD, the amount of reflected light will change according to the reflectivity changes resulting from the hole developing process.   This change in the reflected light is easily detected in a SCOOP.

(A) Static dc measurement

We used a high power GaAlAs laser diode (Sharp Model LT-015 MD) having oscillating wavelength of 780 nm. The numerical aperture of the lens L_1 and L_2 was 0.15 and 0.60 respectively. The focused spot dia-meter on the surface of recording media was about 1.6 μm (1/e^2).   The recording material used in this experiment was a single tellurium alloy film sputtered on a polymethylmethacryate (PMMA) substrate. Two lenses were so set about 20 cm apart as to make the distance between two lenses longer than the coherent length of the LD and thus to eliminate the coherent feedback effect. The LD was drived in the room temper-ature by a constant current source. After making an optical alignments which was considered to realize the maximum output power with an optical feedback at an appropriate injection current region, in other words the minimum threshold current with an optical feedback, we increased gradually the injection currents untill and after the undulations of the output power appeared.   We recorded the signals of injection cur-rents and output powers using an X-Y recorder.

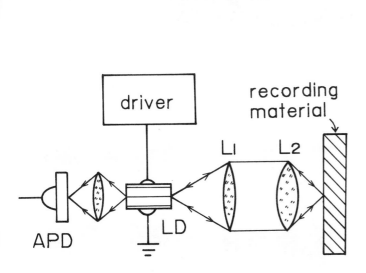

Fig. 8 Experimental setup of a SCOOP for sensitivity measurement of optical storage media.

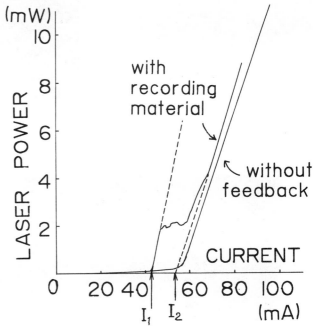

Fig. 9 Output power versus current in a SCOOP with an ablative optical recording material.

Fig. 9 shows the experimental characteristic curves between the injection current and the output power which was calibrated as an absolute power on the surface of the recording material. The threshold current without an optical feedback was about 55 mA. While with an optical feedback it was decreased to about 40 mA. One can note that with an optical feedback resulting from the recording material the curve is undulated around the output power of 2 mW. It is easily presumed that the undulation is originated from the reflectivity changes during the hole formation on the recording material. Thus we can define or determine the static sensitivity of the recording material to be that value. The straight line of the curve above 2 mW output power region is corresponded to the curve with an optical feedback resulting from the decreased reflectivity of the recording material after the hole formation. We note that if we make an auxiliary experiment in order to determine the relations between mirror reflectivities and threshold currents with an optical feedback, we can estimate the reflectivities of the recording material before and after the hole formation using two threshold currents as marked in Fig. 9, I_1 and I_2 respectively. The dotted lines which are extraporated from the straight parts of the curve are equal to those which will be given by an optical feedback resulting from the fixed reflectivity mirror.

(B) Dynamic pulse measurement

In this case we used a modified GaAlAs LD (Hitachi Model HLP 3400), having oscillating wavelength of 820 nm, whose one facet was coated with an antireflection film. The numerical aperture of the collimating lens L_1 and the focusing lens L_2 was 0.3 and 0.6 respectively. The focused spot diameter was about 2 μm ($1/e^2$). The recording material was a tellurium-cupper alloy film sputtered on a PMMA substrate. The LD was drived by a pulsed current source with a proper dc current biased.

Fig. 10 shows the waveforms of (a) the injection current of 40 mA having a duration of about 0.4 μs which was monitored by a current transformer,(b) and (c) were the corresponding output light without and with an optical feedback which were detected by an avalansche photodiode placed at the rear side of the LD. The peak output power without an optical feedback (b) was about 3 mW and that with an optical feedback (c) resulting from an aluminum mirror was about 5 mW. One can note that there are no significant differences in the rising and falling waveform between those without and with an optical feedback. This is partly because the response speed of a SCOOP is quite fast and because the coherent feedback effct is eliminated.

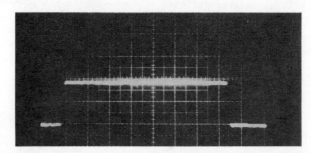

(a) injection current, 50 ns / div, 20 mA / div

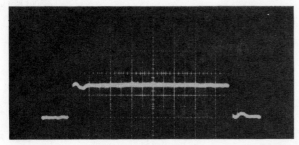 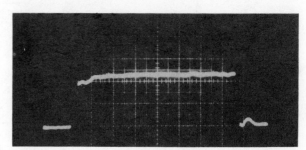

50 ns / div, 2 mW / div.
(b) output light without feedback   (c) output light with feedback
Fig. 10   Pulse waveforms of injection current and output light without and with an optical feedback.

Fig. 11 shows the schematic definition how to determine the sensitivity which is defined by the energy density required to make the reflectivity change in the recording material. First, we make optical alignments as similarly as mentioned above with an appropriate peak power and duration time of the writing laser pulse which will not be able to change the characteristics of recording media. Then gradually we increase the peak power or the duration time of the laser pulse until there will be appear the decrease of the ouput power in the pulse waveform which is schematically depicted in Fig. 11. Second, we measure the writing time which is defined as the time duration as marked in Fig. 11. It is not difficult to do it because there is an abrupt decrease of the output laser power during the hole developing process detected in a SCOOP.

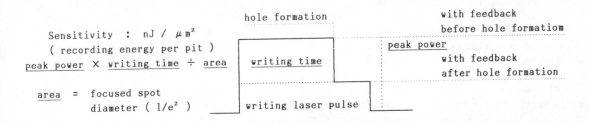

Fig. 11 Definition of sensitivity by energy density.

Then, we can clculate the recording energy multiplying the writing time and the writing peak power whose an absolute value is calibrated separately using a calorimeter. Finally, we determine the recording energy density in the term of nJ / μm² for a pit using the values of focusing spot diameter and recording energy. It is well known that the recording energy will differ from such parameters as duration time, focusing spot diameter, substrate materials etc. In our case, we show the experimental results which are obtained with a fixed focusing spot diameter.

Fig. 12 shows typical experimental waveforms which were obtained with different recording thickness of same optical sensitive materials which are sputtered tellurium-cupper films on a PMMA. One can note that with 8 mW peak recording laser power the writing time for the film with 32 nm thickness was about 0.2 μs while that with 49 nm thickness was about 0.27 μs respectively. These value are quite reasonable that the thicker the film will become, the longer recording time will necessary. The recording energy density itself is almost same with that reported previousely in elsewhere. In Fig. 12 after the abrupt decrease of the output power there are ubdulations in the pulse waveform. They are estimated to be originated partly in the transient phenomena in the recording process, such as vaporization of sensitive materials or diffusion of melted materials.

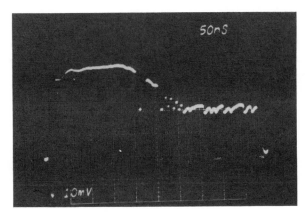
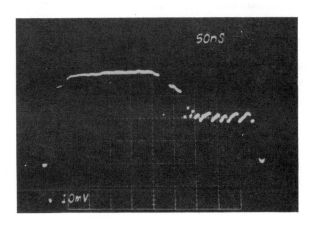

(a) 32 nm thickness   (b) 49 nm thickness

Fig. 12 Transient waveforms of output light in a SCOOP writing an ablative optical recording material.

(C) Evaluation of plastic substrates

Using the experimental setup mentioned above, we evaluated several plastics as a recording substrate in the term of the recording sensitivity. As substrates we examined a PMMA, PC (polycarbonate), PI(polyimide), PVC (polyvinylchrolide) and PET (polyethylenterephthalate). In a vacuum chamber a single tellurium film was deposited on them simultaneously. The thickness of the film was 39 nm. We obtained almost same waveform of a recording laser pulse as shown in Fig. 12 in any cases of different recording materials or duration time of laser light. Results are summarized in Fig. 13, where data are averaged over four times experiments. One can note that there are well known relations where the shorter the writing timer of recording laser pulse is used, the less recording energy density is required. In other words, if we use the higher peak power of the recording laser light, we can use the shorter writing time and thus we can record informations in an ablative optical material with the higher information transfer rate. As for the recording sensitivity of recording media with the different substrates, we can conclude that the medium with a PVC substrate having the recording energy density of about 0.2 nJ / μm^2 is the higest among them. Table 1 shows the required recording energy density at a 0.2 μs writing laser pulse for the media with different plastics and their thermal conductivities. One can note that there are no linear relations between them. Here we do not discuss further for this fact but we explain another experimental results. The two dotted lines in Fig. 13 shows the limits of scattered data for different substrate thickness of a PVC. We prepared the recording media with six different substrate thickness of a PVC varying from 10 to 600 μm. We found that the differences among them in the term of the recording energy density were about ± 0.01 nJ / μm^2 at a 0.2 μs writing laser pulse and they were almost within the measuring error.

In conclusion, we have explained the sensitivity measurement of an optical storage media using a SCOOP and demonstrated its usefulness. It is a great advantage that an LD is used not only to machine holes in an ablative optical recording material but also to monitor the hole development process in real time, the recording and monitoring wavelength of a laser light is the same and the optical alignment of coincident axes in the recording and monitoring beams is automatically fullfilled.

Acknowledgement

We thank Dr. K. Sakurai for his encouragements and persons with Optoelectonics Section for their discussion and Dr. K. Aiki with Hitachi and Dr. T. Hijikata with Sharp for suppling us laser diodes.

References

1. Mitsuhashi,Y., Morikawa,T., Sakurai,K., Seko,A., and Shimada,J., "Self-coupled optical pickup," Optics Commun., Vol. 17, pp. 95-97, 1976

2. Mitsuhashi,Y. and Shimada,J., "Self-coupled optical pickup," Di Chen Chairman/Editor, Proceedings of " Optical Data Storage," January 17-20,1983 Incline Village, Nevada : SPIE Vol. 382, p. 217

3. Haskal,H. M., Di Chen Chairman/Editor, Proceedings of " Optical Data Storage," January 17-20, 1983 Incline Village, Nevada : SPIE Vol. 382, p. 174,
    Lewkowicz,J. and Cohen,D., ibid. p. 236
    Lou, D. Y., ibid. Post Deadline Papers #2
    Blom, G. M., ibid. Post Deadline Papers #3

4. Matsui,S., Takiguchi,H., Hayashi,H., Yamamoto,S., Yano,S. and Hijikata,T., " Suppression of feedback induced noise in short cavity V-channeled substrate inner stripe lasers with self-oscillation," Appl. Phys. Lett. Vol. 43, p. 219, 1983

Table 1 Recording sensitivities of a single tellurium film on various plastic substrates and their thermal conductivities.

Substrate Plastic	Sensitivity (nJ/μ m^2)	Thermal Conductivity (Cal/cm·sec·deg)
P V C (polyvinylchroride)	0.28	4.2 x 10^{-4}
P M M A (polymethylmethacryate)	0.36	4.8 x 10^{-4}
P E (polyethlenterephtalate)	0.39	3.4 x 10^{-4}
P I (polyimide)	0.40	3.7 x 10^{-4}
P C (polycarbonate)	0.47	4.6 x 10^{-4}

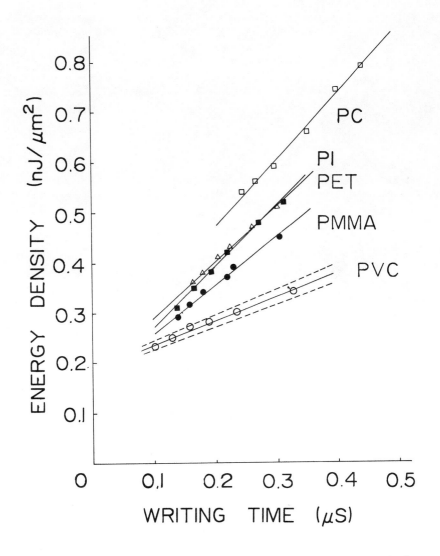

Fig. 13 Dependences of writing time on sensitivity of a single tellurium film recording material with various plastic substrates.

Addendum

The following paper, which was scheduled to be presented at this conference and published in these proceedings, was cancelled.

[490-17] **Future uses of durable ruggedized optical disk data storage devices in spacecraft and mobile platform applications**
B. Thompson, J. Zajaczkowski, Cherokee Data Systems

AUTHOR INDEX